ACTIVE FILTERS
Theory and Design

ACTIVE FILTERS
Theory and Design

S. A. PACTITIS

CRC Press
Taylor & Francis Group
Boca Raton London New York

CRC Press is an imprint of the
Taylor & Francis Group, an **informa** business

CRC Press
Taylor & Francis Group
6000 Broken Sound Parkway NW, Suite 300
Boca Raton, FL 33487-2742

First issued in paperback 2019

© 2008 by Taylor & Francis Group, LLC
CRC Press is an imprint of Taylor & Francis Group, an Informa business

No claim to original U.S. Government works

ISBN-13: 978-1-4200-5476-7 (hbk)
ISBN-13: 978-0-367-38838-6 (pbk)

Library of Congress Cataloging-in-Publication Data

Pactitis, S. A.
    Active filters : theory and design / S.A. Pactitis.
        p. cm.
    Includes bibliographical references and index.
    ISBN 978-1-4200-5476-7 (alk. paper)
    1. Electric filters, Active--Design and construction. I. Title.

TK7872.F5P325 2008
621.3815'324--dc22   2007017798

Visit the Taylor & Francis Web site at
http://www.taylorandfrancis.com

and the CRC Press Web site at
http://www.crcpress.com

Dedication

To the love of my life,
Demetra,
the joy of my life,
Despina, Alexandra, and Maria

Contents

Preface

This book was primarily written to provide readers with a simplified approach to the design of active filters. Filters of some sort are essential to the operation of most electronic circuits. It is therefore in the interest of anyone involved in electronic circuit design to have the ability to design filters capable of meeting a given set of specifications.

Three basic active filter types are used throughout the book: Butterworth, Chebyshev, and Bessel. Those three types of filters are implemented with the Sallen–Key, infinite gain multiple feedback, state-variable, and Biquad circuits that yield low-pass, high-pass, band-pass, and band-reject circuits.

Many examples of low-pass, high-pass, band-pass, and notch active filters are illustrated in complete detail, including frequency normalizing and denormalizing techniques.

It is felt that this book can be used for the following purposes:

1. As a self-study book for practicing engineers and technicians, so that they easily design working filters
2. As a supplementary textbook for graduate or undergraduate courses on the design of filters
3. As a reference book to be used by practicing filter design specialists

S. A. Pactitis

1 Introduction

1.1 FILTERS AND SIGNALS

A **filter** is a circuit that is designed to pass a specified band of frequencies while attenuating all signals outside this band. Filter networks may be either active or passive. *Passive filter networks* contain only resistors, inductors, and capacitors. *Active filters*, which are the only type covered in this text, employ operational amplifiers (op-amps) as well as resistors and capacitors.

The output from most biological measuring systems is generally separable into signal and noise. The signal is that part of the data in which the observer is interested; the rest may be considered noise. This noise includes unwanted biological data and nonbiological interference picked up by or generated in the measuring equipment. Ideally, we would like to remove it while retaining the signal, and often this is possible by suitable filtration.

If the spectra of signal and noise occupy completely separate frequency ranges, then a filter may be used to suppress the noise (Figure 1.1).

As filters are defined by their frequency-domain effects on signals, it makes sense that the most useful analytical and graphical descriptions of filters also fall under the frequency domain. Thus, curves of gain versus frequency and phase versus frequency are commonly used to illustrate filter characteristics, and most widely used mathematical tools are based on the frequency domain.

The frequency-domain behavior of a filter is described mathematically in terms of its **transfer function** or **network function**. This is the ratio of the Laplace transforms of its output and input signals. The voltage transfer function of a filter can therefore be written as

$$H(s) = \frac{V_0(s)}{V_i(s)} \tag{1.1}$$

where s is the complex frequency variable.

The Laplace transform approach to the filter analysis allows the designer to work with algebraic equations in the frequency domain. These are relatively easy to interpret by observation. In contrast, a time-domain approach to filter mathematics results in complex differential equations that are usually far more difficult to manipulate and interpret.

The transfer function defines the filter's response to any arbitrary input signals, but we are most often concerned with its effect on continuous sine waves, especially the magnitude of the transfer function to signals at various frequencies. Knowing the transfer function magnitude (or gain) at each frequency allows us to determine how well the filter can distinguish between signals at different frequencies. The

1

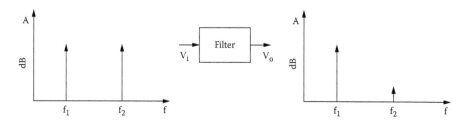

FIGURE 1.1 Using a filter to reduce the effect of an undesired signal.

transfer function magnitude versus frequency is called the **amplitude response** or sometimes, especially in audio applications, the **frequency response**.

Similarly, the **phase response** of the filter gives the amount of **phase shift** introduced in sinusoidal signals as a function of frequency. Because a change in phase of a signal also represents a change in time, the phase characteristics of a filter become especially important when dealing with complex signals in which the time relationships between different frequencies are critical.

By replacing the variables s in equation (1.1) with jw, where $j = \sqrt{-1}$, and w is the radian frequency ($2pf$), we can find the filter's effect on the magnitude and phase of the input signal. The magnitude is found by making the absolute value of Equation (1.1):

$$|H(j\omega)| = \left|\frac{V_0(j\omega)}{V_i(j\omega)}\right| \tag{1.2}$$

or

$$A = 20\log|H(j\omega)| \quad \text{in dB} \tag{1.3}$$

and the phase is

$$\arg H(j\omega) = \arg\frac{V_0(j\omega)}{V_i(j\omega)} \tag{1.4}$$

1.2 BASIC FILTER TYPES

There are four basic filter types:

1. The first type is the **low-pass filter** (LPF). As might be expected, an LPF passes low-frequency signals, and rejects signals at frequencies above the filter's cutoff frequency (Figure 1.2.). The ideal filter has a rectangular shape, indicating that the boundary between the passband and the stopband is abrupt and that the rolloff slope is infinitely steep. This type of response is ideal because it allows us to completely separate signals at different frequencies from one another. Unfortunately, such an amplitude response curve is not physically realizable. We will have to settle for the **approximation** that will still meet our requirements for a given application. Deciding on the best

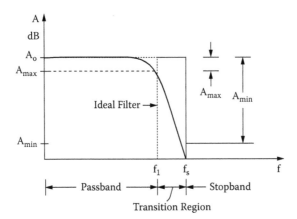

FIGURE 1.2 Low-pass frequency response.

approximation involves making a compromise between various properties of the filter's transfer function. The important properties are the following.

Filter order: The order of a filter has several effects. It is directly related to the number of components in the filter and, therefore, to its price and the complexity of the design task. Therefore, higher-order filters are more expensive, take up more space, and are more difficult to design. The primary advantage of higher-order filters is that they will have steeper rolloff slopes than similar lower-order filters.

Rolloff rate: Usually expressed as the amount of attenuation in dB for a given ratio of frequencies. The most common units are "dB/decade" or "dB/octave."

From Figure 1.2, four parameters are of concern:

A_{max} is the maximum allowable change in gain within the passband. This quantity is also often called the maximum passband ripple.

A_{min} is the minimum allowable attenuation (referred to the maximum passband gain) within the stopband.

f_1 is the cutoff frequency or passband limit.

f_s is the frequency at which the stopband begins.

These four parameters define the order of the filter.

2. The inverse of the low-pass is **the high-pass filter**, which rejects signals below its frequency (Figure 1.3.).

3. **Band-pass filters** pass frequencies within a specified band and reject components outside the band (Figure 1.4).

Band-pass filters are geometrically symmetrical, i.e., symmetrical around a center frequency when plotted on linear-log graph paper with frequency on the logarithmic axis. The center can be computed by:

$$f_0 = \sqrt{f_1 f_2} \qquad (1.5)$$

where f_1 is the lower cutoff and f_2 is the upper cutoff frequency.

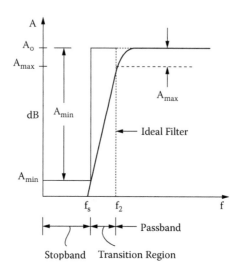

FIGURE 1.3 High-pass frequency response.

For narrow filters, where the ratio of f_2 to f_1 is less than 1.1, the response shape approaches arithmetic symmetry. f_o can then be computed as the average of the cutoff frequencies:

$$f_0 = \frac{f_1 + f_2}{2} \tag{1.6}$$

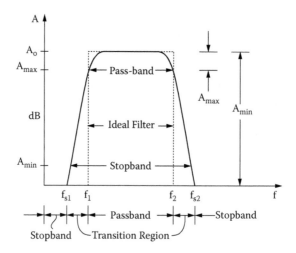

FIGURE 1.4 Band-pass frequency response.

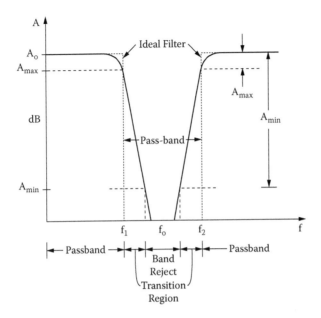

FIGURE 1.5 Band-reject frequency response.

The **selectivity factor** Q is the ratio of the center frequency of a band-pass filter to bandwidth, i.e.,

$$Q = \frac{f_0}{BW} = \frac{f_0}{f_2 - f_1} \tag{1.7}$$

4. **Band-reject** filters reject frequencies within a specified band and pass components outside this band (Figure 1.5).

1.3 THE MATHEMATICS OF ELEMENTARY FILTERS

The transfer functions consist of a numerator divided by a denominator, each of which is a function of s, so they have the form:

$$H(s) = \frac{N(s)}{D(s)} \tag{1.8}$$

The numerator and denominator can always be written as polynomials in s. To be completely general, a transfer function for an nth-order network can be written as

$$H(s) = \frac{K b_0}{s^n + b_{n-1}s^{n-1} + b_{n-2}s^{n-2} + b_1 s + b_0} \tag{1.9}$$

TABLE 1.1

Butterworth filters: $s^n + b_{n-1}s^{n-1} \dots b_1 s + b_0$

n	b_0	b_1	b_2	b_3	b_4	b_5	b_6	b_7
1	1.000							
2	1.000	1.414						
3	1.000	2.000	2.000					
4	1.000	2.613	3.414	2.613				
5	1.000	3.236	5.236	5.236	3.236			
6	1.000	3.863	7.464	9.142	7.464	3.864		
7	1.000	4.494	10.098	14.592	14.592	10.098	4.494	
8	1.000	5.126	13.137	21.846	25.688	21.846	13.137	5.126

where $b_0, b_1, \dots b_{n-1}$ and K are calculated depending on the order of the transfer function n. The Butterworth, Chebyshev, and Bessel filter circuits differ only by the choice of the coefficients b_i, which yield slightly different response curves. The coefficients for the normalized Butterworth, Chebyshev, and Bessel cases are given in Tables 1.1 to 1.7.

Another way of writing a filter's transfer function is to factor the polynomials in the denominator so that they take the form:

$$H(s) = \frac{Kb_0}{(s - p_0)(s - p_1) \dots (s - p_n)} \tag{1.10}$$

The roots of the denominator, $p_0, p_1, \dots, p_n$, are called **poles**. All of the poles will be either real roots or complex conjugate pairs.

Another way to arrange the terms in the network function expression is to recognize that each complex conjugate pair is simply the factored form of a second-order polynomial. By multiplying the complex conjugate pairs out, we can get rid of the complex numbers and put the transfer function into a form that essentially

TABLE 1.2

0.1-dB Chebyshev filter

n	b_0	b_1	b_2	b_3	b_4	b_5	b_6	b_7
1	6.552							
2	3.313	2.372						
3	1.638	2.630	1.939					
4	0.829	2.026	2.627	1.804				
5	0.410	1.436	2.397	2.771	1.744			
6	0.207	0.902	2.048	2.779	2.996	1.712		
7	0.102	0.562	1.483	2.705	3.169	3.184	1.693	
8	0.052	0.326	1.067	2.159	3.419	3.565	3.413	1.681

TABLE 1.3
0.5-dB Chebyshev filter

n	b_0	b_1	b_2	b_3	b_4	b_5	b_6	b_7
1	2.863							
2	1.516	1.426						
3	0.716	1.535	1.253					
4	0.379	1.025	1.717	1.197				
5	0.179	0.753	1.310	1.937	1.172			
6	0.095	0.432	1.172	1.590	2.172	1.159		
7	0.045	0.282	0.756	1.648	1.869	2.413	1.151	
8	0.024	0.153	0.574	1.149	2.184	2.149	2.657	1.146

consists of a number of second-order transfer functions multiplied together, possibly with some first-order terms as well. We can think of the complex filter as being made up of several second-order and first-order filters connected in series. The transfer function thus takes the form:

$$H(s) = \frac{K}{(s^2 + a_{11}s + a_{10})(s^2 + a_{21}s + a_{20})\ldots(s^2 + a_{n1}s + a_{n0})} \tag{1.11}$$

1.3.1 BUTTERWORTH FILTERS

The first, and probably best-known, filter is the **Butterworth** or **maximally flat** response. It exhibits a nearly flat passband. The rolloff is 20 dB/decade or 6 dB/octave for every pole. The general equation for a Butterworth filter's amplitude response is

$$|H(j\omega)| = \frac{K}{\left[1 + \left(\dfrac{s}{\omega_1}\right)^{2n}\right]^{1/2}} \tag{1.12}$$

TABLE 1.4
1-dB Chebyshev filter

n	b_0	b_1	b_2	b_3	b_4	b_5	b_6	b_7
1	1.965							
2	1.103	1.098						
3	0.491	1.238	0.988					
4	0.276	0.743	1.454	0.953				
5	0.123	0.581	0.974	1.689	0.937			
6	0.069	0.307	0.939	1.202	1.931	0.928		
7	0.031	0.214	0.549	1.358	1.429	2.176	0.923	
8	0.017	0.107	0.448	0.847	1.837	1.655	2.423	0.920

TABLE 1.5
2-dB Chebyshev filter

n	b_0	b_1	b_2	b_3	b_4	b_5	b_6	b_7
1	1.308							
2	0.823	0.804						
3	0.327	1.022	0.738					
4	0.206	0.517	1.256	0.716				
5	0.082	0.459	0.693	1.450	0.706			
6	0.051	0.210	0.771	0.867	1.746	0.701		
7	0.020	0.166	0.383	1.144	1.039	1.994	0.698	
8	0.013	0.070	0.360	0.598	1.580	1.212	2.242	0.696

where n is the order of the filter and can be a positive whole number, w_1 is the -3 dB frequency of the filter, and K is the gain of the filter.

We see that $|H(0)| = K$ and $|H(j\omega)|$ is monotonically decreasing with w. In addition, the 0.707 or -3 dB point is at $w = 1$ for all n; that is,

$$|H(j\omega)| = \frac{K}{\sqrt{2}} \quad \text{for all } n \tag{1.13}$$

The cutoff frequency is thus seen to be $w = 1$. The parameter n controls the closeness of approximation in both the band and the stopband.

The amplitude approximation of Equation (1.12) is called **Butterworth** or **maximally flat** response. The reason for the term *maximally flat* is that when we expand $|H(j\omega)|$ in a power series about $w = 0$, we have:

$$|H(j\omega)| = K\left(1 - \frac{1}{2}\omega^{2n} + \frac{3}{8}\omega^{4n} - \frac{5}{16}\omega^{6n} + \frac{35}{128}\omega^{8n} + \cdots\right) \tag{1.14}$$

TABLE 1.6
3-dB Chebyshev filter

n	b_0	b_1	b_2	b_3	b_4	b_5	b_6	b_7
1	1.002							
2	0.708	0.645						
3	0.251	0.928	0.597					
4	0.177	0.405	1.169	0.582				
5	0.063	0.408	0.549	1.415	0.574			
6	0.044	0.163	0.699	0.691	1.663	0.571		
7	0.016	0.146	0.300	1.052	0.831	1.912	0.568	
8	0.011	0.056	0.321	0.472	1.467	0.972	2.161	0.567

TABLE 1.7
Bessel filter

n	b_0	b_1	b_2	b_3	b_4	b_5	b_6
1	1						
2	3	3					
3	15	15	6				
4	105	105	45	10			
5	945	945	420	105	15		
6	10395	10395	4725	1260	210	21	
7	135135	13 5135	62370	17325	3150	378	28

We see that the first $2n-1$ derivatives of $|H(j\omega)|$ are equal to zero at $W = 0$. For $W \gg 1$, the amplitude response of a Butterworth function can be written as (with $K = 1$)

$$|H(j\omega)| \cong \frac{1}{\omega^n} \quad \omega \gg 1 \tag{1.15}$$

We observe that asymptotically, $H(jW)$ falls off as W^{-n} for Butterworth response. In terms of dB, the asymptotic slope is obtained as

$$A = 20\log|H(j\omega)| = -20n\log\omega \tag{1.16}$$

Consequently, the amplitude response falls asymptotically at a rate of $-20n$ dB/decade or $-6n$ dB/octave.

Figure 1.6 shows the amplitude response curves of Butterworth low-pass filters of various orders. The frequency scale is normalized to f/f_1 so that all of the curves show 3-dB attenuation for f/f_1 and $K = 1$.

Figure 1.7 shows the step response of Butterworth low-pass filters of various orders. Note that the amplitude and duration of the ringing increase as n increases.

1.3.2 CHEBYSHEV FILTERS

Another important approximation to the ideal filter is the **Chebyshev** or **equal ripple** response. As the latter name implies, this sort of filter will have a ripple in the passband amplitude response. The amount of passband ripple is one of the parameters used in specifying a Chebyshev filter. The Chebyshev characteristic has a steeper rolloff near the -3 dB frequency when compared to the Butterworth, but at the expense of less "flatness" in the passband and poorer transient response.

The general equation for a Chebyshev filter's amplitude response is

$$|H(j\omega)| = \frac{K}{\left[1+\varepsilon^2 C_n^2(\omega)\right]^{1/2}} \tag{1.17}$$

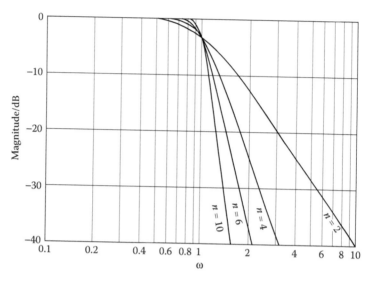

FIGURE 1.6 Amplitude response curves for Butterworth filters of various orders ($K = 1$).

where

$$\cos(n\cos^{-1}\omega) \quad |\omega| \le 1 \tag{1.18}$$

$$C_n(\omega) =$$

$$\cosh(n\cosh^{-1}\omega) \quad |\omega| > 1 \tag{1.19}$$

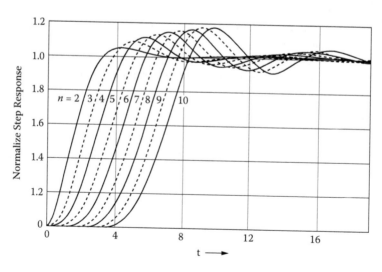

FIGURE 1.7 Step response for Butterworth low-pass filters; $w_1 = 1$ rad/s, and step amplitude is 1.

For $n = 0$ we have:

$$C_0(\omega) = 1 \tag{1.20}$$

and for $n = 1$ we have:

$$C_1(\omega) = \omega \tag{1.21}$$

Higher-order Chebyshev polynomials are obtained through the recursive formula

$$C_n(\omega) = 2\omega\, C_{n-1}(\omega) - C_{n-2}(\omega) \tag{1.22}$$

Thus, for $n = 2$, we obtain $C_2(\omega)$ as

$$C_2(\omega) = 2\omega(\omega) - 1 = 2\omega^2 - 1 \tag{1.23}$$

In Table 1.8 are given Chebyshev polynomials of order up to $n = 8$.

Within the interval $|\omega| \leq 1$, $|H(j\omega)|$ oscillates about unity such that the maximum value is 1 and the minimum is $1/(1+\varepsilon^2)$. Outside this interval, $C_n^2(\omega)$ becomes very large, so that as increases, a point will be reached where $\varepsilon^2 C_n^2(\omega) \gg 1$ and $|H(j\omega)|$ approaches zero very rapidly with further increase in w. Thus, we see that $|H(j\omega)|$ in Equation (1.17) is indeed a suitable approximation for the ideal low-pass characteristics.

Figure 1.8 shows a Chebyshev approximation to the ideal low-pass filter. We see that within the passband $0 \leq \omega \leq 1$, $|H(j\omega)|$ ripples between the value 1 and $1/(1+\varepsilon^2)$. The **ripple height**, or distance between maximum and minimum in the passband, is given as

$$Ripple = 1 - \frac{1}{(1+\varepsilon^2)^{1/2}} \tag{1.24}$$

TABLE 1.8
Chebyshev polynomials $C_n(\omega) = \cos(n \cos^{-1}\omega)$

n	
0	1
1	w
2	$2w^2 - 1$
3	$4w^3 - 3w$
4	$8w^4 - 8w^2 + 1$
5	$16w^5 - 20w^3 + 5w$
6	$32w^6 - 48w^4 + 18w^2 - 1$
7	$64w^7 - 112w^5 + 56w^3 - 7w$
8	$128w^8 - 256w^6 + 160w^4 - 32w^2 + 1$

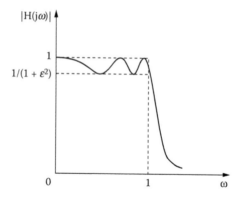

FIGURE 1.8 Chebyshev approximation to low-pass filter.

At W = 1, we have:

$$|H(j1)| = \frac{1}{(1+\varepsilon^2)^{1/2}} \tag{1.25}$$

because $C_n^2(1) = 1$.

In the stopband, that is, for $|\omega \geq 1|$, as ω increases, we reach a point ω_p where $\varepsilon^2 C_n^2(\omega) \gg 1$ so that

$$|H(j\omega)| \cong \frac{1}{\varepsilon C_n(\omega)} \quad \omega > \omega_p \tag{1.26}$$

The transfer function in dB is given as

$$A = 20\log|H(j\omega)|$$

$$\cong -\left[20\log\varepsilon + 20\log C_n(\omega)\right] \tag{1.27}$$

For large ω, $C_n(\omega)$ can be approximated by its leading term $2^{n-1}\omega^n$, so that

$$A = -\left[20\log\varepsilon + 20\log 2^{n-1}\omega^n\right]$$

$$= -\left[20\log\varepsilon + 6(n-1) + 20n\log\omega\right] \tag{1.28}$$

A few different Chebyshev filter responses are shown in Figure 1.9 for various values of n. Note that a filter of order n will have $n-1$ peaks or dips in the

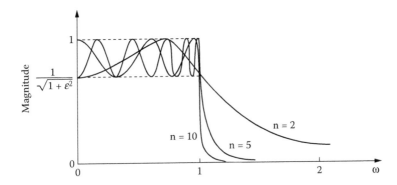

FIGURE 1.9 Examples of Chebyshev amplitude response: (a) 3-dB ripple, (b) expanded view of passband region showing form of responses below cutoff frequency.

passband response. Note also that the nominal gain of the filter ($K = 1$) is equal to the filter's maximum passband gain. An odd-order Chebyshev will have a dc gain to the nominal gain, with "dips" in the amplitude response curve to the ripple value. An even-order Chebyshev will have its dc gain equal to the nominal filter minus the ripple values; the ripple in this case increases the gain to the nominal value.

The addition of a passband ripple as a parameter makes the specification process for a Chebyshev filter more complicated than for a Butterworth filter, but also increases flexibility, because passband ripple can be treated for cutoff slope.

Figure 1.10 shows the step response of 3-dB ripple Chebyshev filters of various orders. As with the Butterworth filters, the higher-order filters ring more.

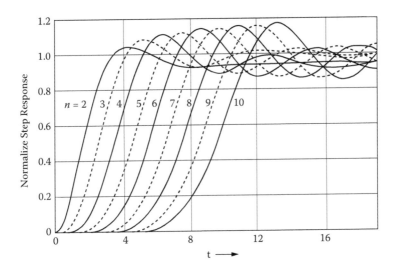

FIGURE 1.10 Step response for Chebyshev low-pass filters; $\omega_1 = 1$ rad/s and step amplitude is 1.

1.3.3 BESSEL–THOMSON FILTERS

So far, filters have been discussed mainly in terms of their amplitude responses, which are plot of gain versus frequency. All these filters exhibit phase shift that varies with frequency. This is an expected and normal characteristic of filters, but in certain cases it can present problems. When a rectangular pulse is passed through a Butterworth or Chebyshev filter, overshoot or ringing will appear on the pulse at the output. If this is undesirable, the Bessel–Thomson filter can be used.

Suppose a system transfer function is given by

$$H(s) = Ke^{-sT} \tag{1.29}$$

where K is a positive real constant. Then, the frequency response of the system can be expressed as

$$H(j\omega) = Ke^{-j\omega T} \tag{1.30}$$

so that the amplitude response $A(\text{w})$ is a constant K, and the phase response

$$\varphi(\omega) = -\omega T \tag{1.31}$$

is **linear** in w. The response of such a system to an excitation is

$$V_0(s) = KV_i(s)e^{-sT} \tag{1.32}$$

so that the inverse transform $v_0(t)$ can be written as

$$v_0(t) = \Im^{-1}\{V_0(s)\} \quad \therefore$$

$$v_0(t) = Kv_i(t-T)u(t-T) \tag{1.33}$$

We see that the response $v_0(t)$ is simply the excitation delayed by a time T, and multiplied by a constant. Thus, no signal distortion results from transmission through a system described by $H(s)$ in Equation (1.29). We note further that the delay T can be obtained by differentiating the phase response $\varphi(\omega)$ by w; that is,

$$delay = -\frac{d\varphi(\omega)}{d\omega} = T \tag{1.34}$$

If ringing or overshoot must be avoided when pulses are filtered, the phase shift between the input and output of a filter must be a linear function of frequency, i.e., the rate of change of the phase with respect to frequency must be constant. The net effect of a constant group delay in a filter is that all frequency components of a signal transmitted through it are delayed by the same amount, i.e., there is no dispersion of signals passing through the filter. Accordingly, because a pulse contains signals of different frequencies, no dispersion takes place, i.e., its shape will be retained when it is filtered by a network that has a linear phase response or constant group delay.

Just as the Butterworth filter is the best approximation to the ideal of "perfect flatness of the amplitude response" in the filter passband, so the Bessel filter provides the best approximation to the ideal of "perfect flatness of the group delay" in the passband, because it has a maximally flat group delay response. However, this applies only to low-pass filters because high-pass and band-pass Bessel filters do not have the linear-phase property. Figure 1.11 compares the amplitude (a) and phase response

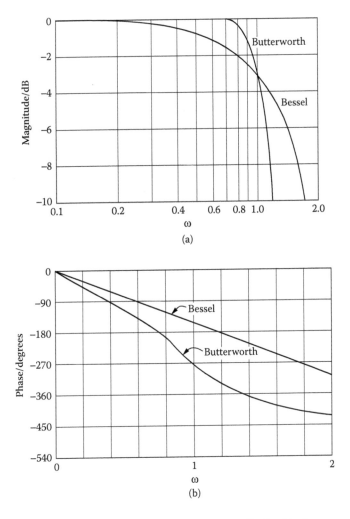

FIGURE 1.11 Response of Butterworth and Bessel: (a) amplitude, (b) phase.

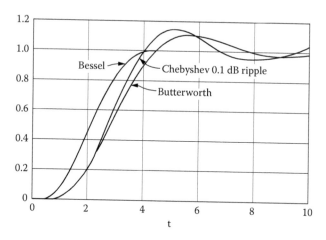

FIGURE 1.12 Step response for Bessel low-pass filters; $w_1 = 1$ rad/s and input step amplitude is 1.

of a Bessel filter with that of a Butterworth filter of the same order. Bessel step response is plotted in Figure 1.12 for various values of n.

To determine which basic filter is most suitable for a given application, it is useful to have a side-by-side comparison of their amplitude, phase, and delay characteristics. These characteristics are determined by the location, in the s plane, of the n poles of $H(s)$. Thus, the poles of a Butterworth filter lie on a semicircle in the left-half s plane, those of a Chebyshev filter on an ellipse that becomes narrower with increasing ripple, and those of a Bessel filter on a curve outside the Butterworth semicircle. This is shown in Figure 1.13.

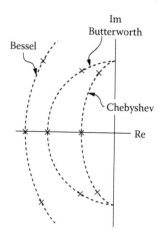

FIGURE 1.13 Comparison of pole location of Butterworth, Chebyshev, and Bessel filters.

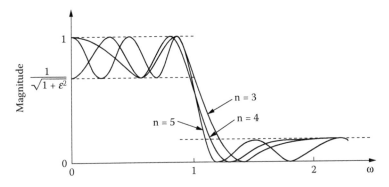

FIGURE 1.14 Magnitude characteristics of an elliptic LP with arbitrary ε and $n = 2, 4$, and 5.

1.3.4 ELLIPTIC OR CAUER FILTERS

Cauer filters, or *elliptic filters,* have ripples both in the passband and in the stopband for an even sharper characteristic in the transition band. Consequently, they can provide a given transition-band cutoff rate with an even lower-order n than Chebyshev filters.

The elliptic filters give a sharp cutoff by adding notches in the stopband. These cause the transfer function to drop at one or more frequencies in the stopband Figure 1.14.

The squared magnitude response is given by

$$| H(j\Omega) |^2 = \frac{1}{1 + \varepsilon^2 U_n^2 (\Omega, L)} \tag{1.35}$$

where ε is the passband parameter and $U_n(\Omega, L)$ is the nth-order Jacobian elliptic function. The parameter L contains information about relative heights of ripples in the passband and stopband.

1.4 WHY ACTIVE FILTERS?

An **active filter** is a network of passive R, C elements, and one or more active elements. Its function is to simulate the action of the usual passive RLC filters. The active element is usually one or more op-amps.

The single system parameter that dictates the filter technology is frequency. Figure 1.15 illustrates the advantages of active filtering compared to passive techniques as a function of frequency.

Active filters offer accuracy, stable tuning, and high immunity to electromagnetic interference. The high input and low output impedance found in active filters allow combinations of two or more stages without the interaction found in passive cascades.

Active filters function similar to simple, frequency-selective control systems; as such, any desired filter characteristic can be generated from the interconnection of

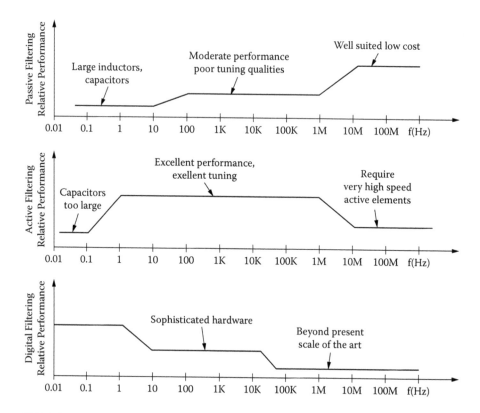

FIGURE 1.15 Comparing filter techniques.

integrators, inverter, summing amplifier, and lossy integrators. Efficient and low-cost active filter design, therefore, depends on the realization of a desired transfer function into a circuit that uses the fewest components while maintaining all performance requirements. Numerous circuits have evolved to meet this objective. Some of the more common are Sallen–Key, multifeedback, state-variable, and biquad. Each circuit has been designed to optimize specific performance aspects. Some are easily tuned, others use a minimum number of components, and still others feature a fixed bandwidth. Most designers would rather not "reinvent the wheel"; however, they would like to develop a basic engineering understanding of the operation, advantages, and restrictions of each type of filter design.

Once the concept of the active filter as a frequency-selective control system is understood, active filter analysis becomes straightforward. In the next chapters we will present some basic active filter techniques. Several examples will be used to show the trade-offs and guidelines of active filter design and op-amp selection.

1.5 PRACTICAL APPLICATIONS

Active filters find use in a wide variety of applications. Some of the more widely recognized are described below.

1.5.1 TONE SIGNALING

The "touch-tone" telephone systems use active filters to decode the dual tone generated at the telephone into the characters 0–9, "*", and "#".

1.5.2 BIOFEEDBACK

The four commonly recognized brainwaves, delta, theta, alpha, and beta, can be segregated using active band-pass filtering techniques. Various graphical techniques may be employed to monitor transitions between wave-ground as a response to stimuli. Careful attention must be paid to the amplifier's low-frequency noise characteristic. The OP-07 (PMI) has the lowest noise of any monolithic op-amp.

1.5.3 INSTRUMENTATION

Applications of active filters to instrumentation is the most diverse of all fields. The low-pass filter is often used as a signal conditioner or noise filter. Because low-pass filters are designated to operate down to dc, low op-amp offset voltages, such as those offered by the OP-07, are of prime importance. In harmonic distortion measurements, notch or bandstop filters allow harmonic interference to be accurately determined.

1.5.4 DATA ACQUISITION SYSTEMS

The noise caused by input switches or high-speed logic is removed with low-pass active filters. Signals reproduced from digital information through a digital-to-analog converter often appear as staircases. In more extreme cases, because of limited sampling, the reproduction appears at only a few discrete levels; however, sophisticated filters can accurately reconstruct the input signal. Remote sensing, in noisy environments, requires the noise-rejecting properties of simple active filters.

1.5.5 AUDIO

Electronic music and audio equalizers use a large number of active filters. Synthesizers combine various low-pass, high-pass, and bandpass functions to generate waveforms that have spectral densities similar to orchestra instruments. Symmetrical positive and negative slew rates make the OP-11 (PMI) well suited for audio applications.

1.5.6 LAB SIGNAL SOURCES

Because of the active filter characteristics, high-purity oscillators are easily designed with very few components.

1.6 THE VOLTAGE-CONTROLLED VOLTAGE SOURCE (VCVS)

In many applications where a high-impedance source must be interfaced to a filter, a noninverting VCVS op-amp filter may be used. Typical input impedances are greater than tens of megaohms, and output impedances are typically less than a few ohms. This, of course, depends on the amplifier being used.

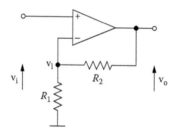

FIGURE 1.16 A noninverting VCVS operational amplifier with resistive feedback.

For a circuit operating in the noninverting mode, for an ideal op-amp, we find from figure 1.16, the gain K.

From node v_1, we have:

$$-G_2 V_0 + (G_1 + G_2)V_1 = 0 \quad \therefore$$

$$V_0 = \left(1 + \frac{G_1}{G_2}\right)V_i \quad \therefore$$

$$V_0 = \left(1 + \frac{R_2}{R_1}\right)V_i \quad \therefore$$

$$K = \frac{V_0}{V_1} = 1 + \frac{R_2}{R_1} \qquad (1.36)$$

because $V_i = V_1$ (for an ideal op-amp).

Generally speaking, the VCVS active filters are much easier to tune and are adjustable over a wider range, without affecting the network parameters, than the infinite gain topologies.

2 Sallen–Key Filters

2.1 INTRODUCTION

There are many ways of constructing active filters. One general-purpose circuit that is widely used is that of Sallen and Key. We refer to the Sallen and Key circuit as a VCVS because it uses an op-amp and two resistors connected so as to constitute a voltage-controlled voltage source (VCVS). Such a configuration offers good stability, requires a minimum number of elements, and has low impedance, which is important for cascading filters with four or more poles.

2.2 FREQUENCY RESPONSE NORMALIZATION

Several parameters are used to characterize a filter's performance. The most commonly specified parameter is frequency response. When given a frequency-response specification, the designer must select a filter design that meets these requirements. This is accomplished by transforming the required response to a normalized low-pass specification having a cutoff of 1 rad/s. This normalized response is compared with curves of normalized low-pass filters that also have a 1 rad/s cutoff. After a satisfactory low-pass filter is determined from the curves, the tabulated normalized element values of the chosen filter are transformed or denormalized to the final design.

The basic for normalization of filters is the fact that a given filter's response can be scaled or shifted to a different frequency range by dividing the reactive elements by a frequency-scaling factor (FSF). The FSF is the ratio of the desired cutoff frequency of the active filter to the normalized cutoff frequency, i.e.:

$$FSF = \frac{\omega_1}{\omega_n} = \frac{2\pi f_1}{1} = 2\pi f_1 \qquad (2.1)$$

The FSF must be a dimensionless number. So, both the numerator and denominator of Equation (2.1) must be expressed in the same units, usually rad/s.

Frequency-scaling a filter has the effect of multiplying all points on the frequency axis of the response curve by the FSF. Therefore, a normalized response curve can be directly used to predict the attenuation of the denormalized filter.

Any linear active or passive network maintains its transfer function if all resistors are multiplied by an impedance-scaling factor (ISF) and all capacitors are divided by the same factor ISF. This occurs because the ISFs cancel one another out in the transfer function. Impedance scaling can be mathematically expressed as

$$R = ISF \times R_n \tag{2.2}$$

$$C = \frac{C_n}{ISF} \tag{2.3}$$

where R_n and C_n are the normalized values.

Frequency and impedance scaling are normally combined into one step rather than performed sequentially. The denormalized values are then given by

$$R = ISF \times R_n \tag{2.4}$$

$$C = \frac{C_n}{ISF \times FSF} \tag{2.5}$$

2.3 FIRST-ORDER LOW-PASS FILTER

Figure 2.1 shows the first-order low-pass filter with noninverting gain K.

From this figure, we have:

node v_1

$$-GV_i + (G + sC)V_1(s) = 0 \quad \therefore$$

$$V_1(s) = \frac{G}{G + sC}V_i$$

where

$$V_0 = KV_1 \quad \text{and} \quad K = 1 + \frac{R_b}{R_a} \quad \backslash$$

$$V_0(s) = \frac{KG}{G + sC}V_i \quad \therefore$$

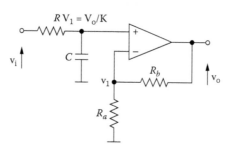

FIGURE 2.1 First-order LPF with gain K.

$$H(s) = \frac{V_0(s)}{V_i(s)} = \frac{KG}{G + sC}$$

$$H(s) = \frac{K\omega_1}{s + \omega_1} = \frac{K}{1 + \dfrac{s}{\omega_1}} \tag{2.6}$$

where

$$s = j\omega$$

$$\text{and} \quad b = \omega_1 = \frac{1}{RC} \tag{2.7}$$

$\omega_1 = 2\pi f_1$, f_1 is the cutoff frequency of the filter. For the normalized filter $G_n = 1\,S\ (R_n = 1\,\Omega)$.
    Hence

$$C_n = \frac{1}{b} \tag{2.8}$$

For $R_{an} = 1\,\Omega \qquad \therefore$

$$R_{bn} = K - 1\,\Omega \tag{2.9}$$

2.3.1 FREQUENCY RESPONSE

From the transfer function, Equation (2.6), we have:

1. For $\dfrac{s}{\omega_1} \ll 1$, we have:

$$H(j\omega) \cong K$$

    The slope is 0 dB/dec and

$$A = 20\log|H(j\omega)| = 20\log K \text{ dB}$$

2. For $\dfrac{s}{\omega_1} \gg 1 \qquad \therefore$

$$|H(j\omega)| = \frac{K}{\dfrac{\omega}{\omega_1}} = K\left(\frac{\omega}{\omega_1}\right)^{-1} \qquad \therefore$$

$$A = 20\log|H(j\omega)| = 20\log K + 20\log\left(\frac{\omega}{\omega_1}\right)^{-1} = 20\log K - 20\log\left(\frac{\omega}{\omega_1}\right) \text{ dB}$$

For $\dfrac{\omega}{\omega_1} = 10$ \ $slope = -20$ dB/dec

For $\dfrac{\omega}{\omega_1} = 2$ $\quad\therefore\quad$ $slope = -6$ dB/oct

$A = 20 \log K - 20 \log 10 \text{ dB} = 20 \log K - 20 \text{ dB}$

3. For $\dfrac{s}{\omega_1} = 1$ $\quad\therefore$

$$H(j\omega) = \frac{K}{1+j} \qquad \therefore$$

$$\left|H(j\omega)\right| = \frac{K}{\sqrt{2}} \qquad \therefore$$

$$A = 20 \log \frac{K}{\sqrt{2}} = 20 \log K - 20 \log \sqrt{2} = 20 \log K - 3 \text{ dB}$$

Figure 2.2 shows the frequency response of the filter.

EXAMPLE 2.1

A first-order LP Butterworth filter must be designed with gain of 5 at a cutoff frequency of 1 kHz.

Solution

From the Butterworth coefficients of Appendix C, we have ($n = 1$):
  $b = 1$, hence, from Equation (2.9), we have:

$$C_n = 1\text{F}$$

$$R_n = 1\,\Omega, \quad R_{an} = 1\,\Omega, \quad \text{and} \quad R_{bn} = K - 1 = 5 - 1 = 4\,\Omega$$

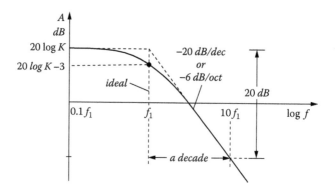

FIGURE 2.2 Frequency response of first-order LPF.

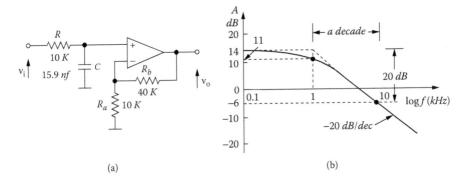

FIGURE 2.3 LP Butterworth filter, where $f_1 = 1$ kHz and $K = 5$ (14 dB).

Denormalization

$$ISF = 10^4$$

$$FSF = \frac{\omega_1}{\omega_n} = \frac{2\pi f_1}{1} = 2\pi \times 10^3 \quad \therefore$$

$$C = \frac{C_n}{ISF \times FSF} = \frac{1}{2\pi \times 10^7} = 15.9 \text{ nF}$$

$$R = ISF \times R_n = 10^4 \times 1 \, \Omega = 10 \text{ k}\Omega$$

$$R_a = ISF \times R_{an} = 10 \times 1 \text{ k}\Omega = 10 \text{ k}\Omega$$

$$R_b = ISF \times R_{bn} = 10 \times 4 \text{ k}\Omega = 40 \text{ k}\Omega$$

Figure 2.3 shows the designed filter with its frequency response.

2.4 FIRST-ORDER HIGH-PASS FILTER

Figure 2.4 shows the first-order high-pass filter with noninverting gain K.

$$K = 1 + \frac{R_b}{R_a}$$

For the node v_0/K, we have:

$$-sCV_i + (G + sC)\frac{V_0}{K} = 0 \quad \therefore$$

$$H(s) = \frac{V_0}{V_i} = \frac{KsC}{G + sC} = \frac{Ks}{s + \dfrac{1}{RC}} = \frac{Ks}{s + \omega_2} \tag{2.10}$$

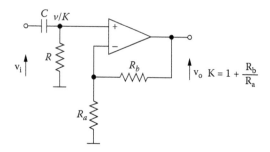

FIGURE 2.4 First-order HPF with gain K.

where

$$\omega_2 = \frac{1}{RC} = b \tag{2.11}$$

For the normalized filter $C_n = 1F$, $\omega_{2n} = 1\,\text{rad/s}$, hence

$$R_n = \frac{1}{b} \tag{2.12}$$

and for $R_{an} = 1\,\Omega$ $\therefore$

$$R_{bn} = K - 1 \tag{2.13}$$

2.4.1 FREQUENCY RESPONSE

From Equation (2.10), we have:

$$H(s) = \frac{K\left(\dfrac{s}{\omega_2}\right)}{1 + \left(\dfrac{s}{\omega_2}\right)} \tag{2.14}$$

From this transfer function, we have:

1. For $\dfrac{s}{\omega_2} \ll 1$ $\therefore$

$$H(j\omega) = K\left(j\frac{\omega}{\omega_2}\right) \qquad \therefore$$

$$A = 20\log|H(j\omega)| = 20\log K + 20\log\left(\frac{\omega}{\omega_2}\right)\,\text{dB}$$

For $\dfrac{\omega}{\omega_2} = 10$ $\therefore$ *slope* = 20 dB/dec.

For $\dfrac{\omega}{\omega_2} = 2$ $\therefore$ *slope* = 6 dB/oct. $\therefore$

$$A = 20 \log K + 20 \text{ dB}$$

2. For $\dfrac{s}{\omega_2} \gg 1$ $\therefore$ $H(j\omega) = K$ \

$$A = 20 \log |H(j\omega)| = 20 \log K \text{ dB}$$

The slope is 0 dB/dec.

3. For $\dfrac{s}{\omega_2} = 1$ $\therefore$

$$H(j\omega) = \frac{K}{1+j} \quad \therefore \quad |H(j\omega)| = \frac{K}{\sqrt{2}} \quad \therefore$$

$$A = 20 \log \frac{K}{\sqrt{2}} = 20 \log K - 20 \log \sqrt{2} = 20 \log K - 3 \text{ dB}$$

Figure 2.5 shows the frequency response of the filter.

EXAMPLE 2.2

A first-order HP Butterworth filter must be designed with gain of 5 at a cutoff frequency of 100 Hz.

Solution

From the Butterworth coefficients of Appendix C, we have ($n = 1$):
$b = 1$, hence:

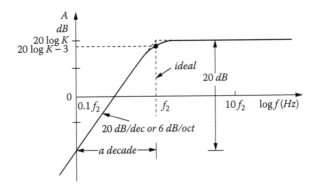

FIGURE 2.5 Frequency response of first-order HPF.

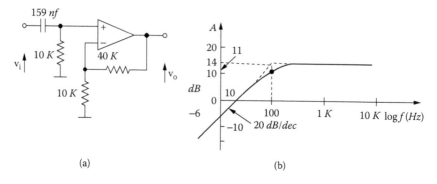

FIGURE 2.6 HP Butterworth filter, where $f_2 = 100$ Hz and $K = 5$ (14 dB).

$$C_n = 1 \text{ F} \qquad R_n = 1 \, \Omega$$

$$R_{an} = 1 \, \Omega \quad \text{and} \quad R_{bn} = K - 1 = 4 \, \Omega$$

Denormalization

$$ISF = 10^4$$

$$FSF = \frac{\omega_2}{\omega_n} = \frac{2\pi f_2}{1} = 2\pi \times 100 = 200\pi \qquad \backslash$$

$$C = \frac{C_n}{ISF \times FSF} = \frac{1}{2\pi \times 10^6} = 159.2 \text{ nF}$$

$$R = ISF \times R_n = 10^4 \times 1 \, \Omega = 10 \text{ k}\Omega$$

$$R_b = ISF \times R_{bn} = 10 \times 1 \text{ k}\Omega$$

$$R_b = ISF \times R_{bn} = 10 \times 4 \text{ k}\Omega = 40 \text{ k}\Omega$$

Figure 2.6 shows the designed filter with its frequency response.

2.5 SECOND-ORDER FILTERS

In a passive RC network (Figure 2.7), the transfer function has poles that lie solely on the negative real axis of the complex-frequency plane.

For the active elements, an ideal voltage amplifier can be used. And because the device is ideal, its characteristics are a gain of K, zero phase shift, infinite input impedance, and zero output impedance (Figure 2.7b). Adding the amplifier to the

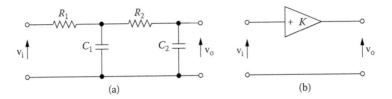

(a)   (b)

FIGURE 2.7 (a) Passive RC network, (b) active device (op-amp).

RC network is best achieved in a feedback configuration, although there are several ways to choose (Figure 2.8a).

From Figure 2.8a, for node v_1, we have:

$$-\frac{1}{R_1}V_i + \left(\frac{1}{R_1} + \frac{1}{R_2} + sC_1\right)V_1 - \frac{1}{R_2}V_2 - sC_1V_0 = 0 \qquad (2.15)$$

node v_2

$$-\frac{1}{R_2}V_1 + \left(\frac{1}{R_2} + sC_2\right)V_2 = 0 \qquad (2.16)$$

but

$$V_2 = \frac{V_0}{K} \qquad (2.17)$$

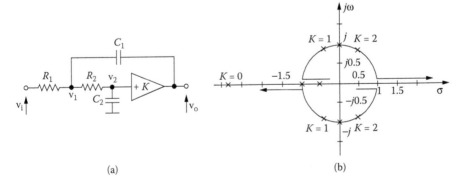

(a)   (b)

FIGURE 2.8 (a) An RC passive network and an op-amp are combined to form a general active filter. (b) When the gain K is varied throughout the s-plane, the response with respect to pole position is obtained.

From Equations (2.16) and (2.17), we have:

$$\frac{1}{R_2}V_1 = \left(\frac{1}{R_2} + sC_2\right)\frac{V_0}{K} \qquad \therefore$$

$$V_1 = (1 + sR_2C_2)\frac{V_0}{K} \tag{2.18}$$

From Equations (2.15), (2.17), and (2.18), we have:

$$(R_1 + R_2 + sR_1R_2C_1)V_1 - R_1\frac{V_0}{K} - sR_1R_2C_1V_0 = R_2V_i \qquad \therefore$$

$$(1 + sR_2C_2)(R_1 + R_2 + sR_1R_2C_1)\frac{V_0}{K} - R_1\frac{V_0}{K} - sR_1R_2C_1V_0 = R_2V_i \qquad \therefore$$

$$R_2[s^2R_1R_2C_1C_2 + s(R_1C_1 + R_1C_2 + R_2C_2 - R_1C_1K) + 1]V_0 = KR_2V_i \qquad \therefore$$

$$H(s) = \frac{V_0}{V_i} = \frac{K}{s^2R_1R_2C_1C_2 + s[C_2(R_1 + R_2) + R_1C_1(1 - K)] + 1}$$

$$H(s) = \frac{K}{as^2 + bs + 1} \tag{2.19}$$

where

$$a = R_1R_2C_1C_2 \qquad \text{and} \qquad b = C_2(R_1 + R_2) + R_1C_1(1 - K) \tag{2.20}$$

which reduces to the transfer function of the RC network itself when $K = 0$. The poles are

$$s = \frac{-b \pm \sqrt{b^2 - 4ac}}{2a}$$

where

$$a = C_1C_2R_1R_2, \qquad b = C_1R_1(1 - K) + C_2(R_1 + R_2), \qquad c = 1$$

With $C_1 = 1/\sqrt{2}$, $C_2 = \sqrt{2}/2$, $R_1 = R_2 = 1$, and $K = 0$, the poles are on the negative real axis, one pole at -2.41 and the other at -0.414 (Figure 2.8b). As the value of K increases, the poles move toward each other until, at $K = 0.586$, both poles converge at -1. As K is increased further, the poles go different ways and follow the circular paths of unit radius. At $K = 2$, the poles cross the jw axis at $\pm j1$. The behavior in the right half-plane for larger values of K is then analogous to that in the left half-plane, with the right half-plane characteristics representing unstable network behavior.

Of interest are the network characteristics at $K = 1$. In this case, the poles are located at $s = -0.707 \pm j0.707$, which is the same as that of a two-pole Butterworth low-pass filter.

2.6 LOW-PASS FILTERS

The basic circuit of the VCVS low-pass filter is shown in Figure 2.9. From this figure, we have:

$$K = 1 + \frac{R_b}{R_a}$$

node v_1

$$-GV_i + (G_1 + G_2 + sC_2)V_1 - G_2 \frac{V_0}{K} - sC_2V_0 = 0 \tag{2.21}$$

node $v_2(v_0/K)$

$$-G_2V_1 + (G_2 + sC_1)\frac{V_0}{K} = 0 \tag{2.22}$$

\

$$V_1 = \frac{(G_2 + sC_1)V_0}{KG_2} \tag{2.23}$$

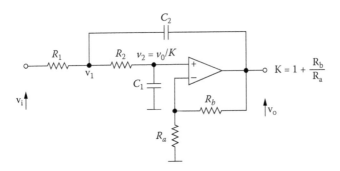

FIGURE 2.9 A second-order Sallen–Key (VCVS) LPF.

From Equations (2.21) and (2.23), we have:

$$\frac{(G_2 + sC_1)(G_1 + G_2 + sC_2)}{KG_2}V_0 - \frac{G_2}{K}V_0 - sC_2V_0 = G_1V_i \qquad \therefore$$

$$\{s^2C_1C_2 + s[C_1(G_1 + G_2) + (1 - K)C_2G_2] + G_1G_2\}V_0 = KG_1G_2V_i \qquad \therefore$$

$$H(s) = \frac{K\dfrac{G_1G_2}{C_1C_2}}{s^2 + s\left[\dfrac{G_1 + G_2}{C_2} + \dfrac{(1 - K)G_2}{C_1}\right] + \dfrac{G_1G_2}{C_1C_2}} \tag{2.24}$$

$$H(s) = \frac{Kb}{s^2 + as + b} \tag{2.25}$$

where

$$\omega_1^2 = b = \frac{G_1G_2}{C_1C_2} \tag{2.26}$$

and

$$a = \frac{G_1 + G_2}{C_2} + \frac{(1 - K)G_2}{C_1} \tag{2.27}$$

For $R_{1n} = R_{2n} = R_n = 1\,\Omega \qquad \therefore$

$$b = \frac{1}{C_1C_2} \tag{2.28}$$

and

$$a = \frac{2}{C_2} + \frac{1 - K}{C_1} \tag{2.29}$$

2.6.1 FREQUENCY RESPONSE

From Equations (2.24) and (2.26), we have:

$$H(s) = \frac{K\omega_1^2}{s^2 + s\left[\dfrac{G_1 + G_2}{C_2} + \dfrac{(1 - K)G_2}{C_1}\right] + \omega_1^2} \qquad \therefore$$

$$H(s) = \frac{K}{\left(\dfrac{s}{\omega_1}\right)^2 + \dfrac{1}{\omega_1}\left[\dfrac{G_1 + G_2}{C_2} + \dfrac{(1 - K)G_2}{C_2}\right]\dfrac{s}{\omega_1} + 1} \qquad \therefore$$

$$H(s) = \frac{K}{\left(\dfrac{s}{\omega_1}\right)^2 + a\left(\dfrac{s}{\omega_1}\right) + 1} \tag{2.30}$$

1. For $\dfrac{s}{\omega_1} \ll 1$, we have:

$$H(j\omega) \cong K$$

The slope is 0 dB/dec and

$$A = 20\log|H(j\omega)| = 20\log K \text{ dB}$$

2. For $\dfrac{s}{\omega_1} \gg 1$, we have:

$$|H(j\omega)| \cong \frac{K}{\left(\dfrac{\omega}{\omega_1}\right)^2} = K\left(\frac{\omega}{\omega_1}\right)^{-2} \quad \therefore$$

$$A = 20\log|H(j\omega)| = 20\log K\left(\frac{\omega}{\omega_1}\right)^{-2} = 20\log K - 40\log\left(\frac{\omega}{\omega_1}\right) \text{ dB}$$

For $\dfrac{\omega}{\omega_1} = 10 \quad \therefore \quad slope = -40 \text{ dB/dec}$

For $\dfrac{\omega}{\omega_1} = 2 \quad \therefore \quad slope = -12 \text{ dB/oct}$

$$A = 20\log K - 40 \text{ dB}$$

3. For $\dfrac{s}{\omega_1} = 1 \quad \therefore$

$$H(j\omega) = \frac{K}{1+j} \quad \therefore \quad |H(j\omega)| = \frac{K}{\sqrt{2}}$$

$$A = 20\log\frac{K}{\sqrt{2}} = 20\log K - 20\log\sqrt{2} = 20\log K - 3 \text{ dB}$$

Figure 2.10 shows the frequency response of the filter.

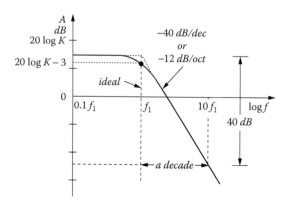

FIGURE 2.10 Frequency response of second-order LPF.

2.6.2 DESIGN PROCEDURE

First method
From Equation (2.28), we have:

$$C_2 = \frac{1}{bC_1}$$

(2.31)

From Equations (2.30) and (2.29), we have:

$$2bC_1 + \frac{1-K}{C_1} = a \quad \therefore \quad 2bC_1^2 - aC_1 + 1 - K = 0 \quad \therefore$$

$$C_1 = \frac{a \pm \sqrt{a^2 - 8b(K-1)}}{4b}$$

(2.32)

The capacitance C_1 could have two values that satisfy Equation (2.31) because plus and minus signs appear outside the radical. However, we will use the plus sign on the radical for our solutions to active filter design, hence:

$$C_{1n} = \frac{a + \sqrt{a^2 + 8b(K-1)}}{4b}$$

(2.33)

and from Equations (2.31) and (2.32), we have:

$$C_{2n} = \frac{4}{a + \sqrt{a^2 + 8b(K-1)}}$$

(2.34)

The radical in the foregoing equations must be positive, hence:

$$K \geq 1 - \frac{a^2}{4b} \tag{2.35}$$

For $K = 1$, Equation (2.33) becomes:

$$C_{1n} = \frac{a}{2b} \tag{2.36}$$

and from Equations (2.34) and (2.31), we have:

$$C_{2n} = \frac{2}{a} \tag{2.37}$$

EXAMPLE 2.3

A second-order low-pass Butterworth filter must be designed with gain 10 and $f_1 = 1$ kHz.

Solution

From Butterworth coefficients in Appendix C, for $n = 2$, we have:

$$a = 1.414, \, b = 1.000$$

From Equation (2.33), we have:

$$C_1 = \frac{1.414 + \sqrt{1.414^2 + 8 \times 9}}{4} = \frac{1.414 + 8.602}{4} = 2.504 \, \text{F}$$

and from Equation (2.31):

$$C_2 = \frac{1}{2.504} = 0.399 \, \text{F}$$

Finally, we denormalize the resistance and capacitance values. We find the frequency-normalizing factor FSF and the impedance-scaling factor ISF from Equations (2.4) and (2.5), respectively.

$$ISF = 10^4 \, \text{ and}$$

$$FSF = \frac{\omega_1}{\omega_n} = \frac{2\pi f_1}{1} = 2\pi \times 10^3 \quad \therefore$$

$$R = ISF \times R_n = 10^4 \times 1\,\Omega = 10\ \text{k}\Omega$$

$$C_1 = \frac{C_{1n}}{ISF \times FSF} = \frac{2.504}{2\pi \times 10^7} \qquad \therefore \qquad C_1 = 39.9\ \text{nF}$$

$$C_2 = \frac{C_{2n}}{ISF \times FSF} = \frac{0.399}{2\pi \times 10^7} \qquad \therefore \qquad C_2 = 6.4\ \text{nF}$$

$$R_b = (K-1)R_a$$

For $\quad R_a = 1\ \text{k}\Omega \quad \therefore \quad R_b = 9R_a \quad$ or $\quad R_b = 9\ \text{k}\Omega$

The active filter circuit designed in the example is shown in Figure 2.11a and its frequency response in Figure 2.11b.

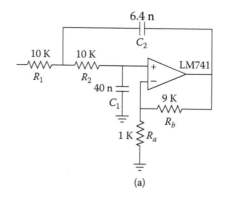

(a)

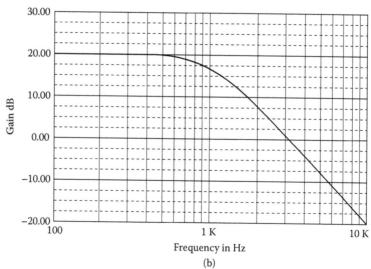

(b)

FIGURE 2.11 (a) The second-order Butterworth VCVS LPF, $f_1 = 1$ kHz, $K = 10$; (b) its frequency response.

EXAMPLE **2.4**

A second-order LP Butterworth filter must be designed with gain 1 and $f_1 = 750$ Hz.

Solution

From the Butterworth coefficients in Appendix C, for $n = 2$, we have:

$$a = 1.414, \quad b = 1.000$$

From Equations (2.36) and (2.37) we have, respectively:

$$C_1 = \frac{a}{2b} = \frac{1.414}{2} = 0.707$$

$$C_2 = \frac{2}{a} = \frac{2}{1.414} = 1.414$$

Denormalization

$$ISF = 10^4$$

$$FSF = \frac{\omega_1}{\omega_n} = \frac{2\pi f_1}{1} = 2\pi \times 750 = 1.5\pi \times 10^3$$

$$C_1 = \frac{C_{1n}}{ISF \times FSF} = \frac{0.707}{1.5\pi \times 10^7} = 15 \text{ nF}$$

$$C_2 = \frac{C_{2n}}{ISF \times FSF} = \frac{1.414}{1.5\pi \times 10^7} = 30 \text{ nF}$$

$$R = ISF \times R_n = 10^4 \times 1 \ \Omega = 10 \text{ k}\Omega$$

The final circuit looks like the one shown below in Figure 2.12a, with its frequency response in Figure 2.12b.

EXAMPLE **2.5**

Design a 300-Hz LP Chebyshev filter with gain 5 and ripple width 3 dB.

Solution

From Chebyshev 3-dB coefficients in Appendix C, we find:

$$a = 0.645, \ b = 0.708$$

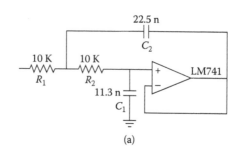

(a)

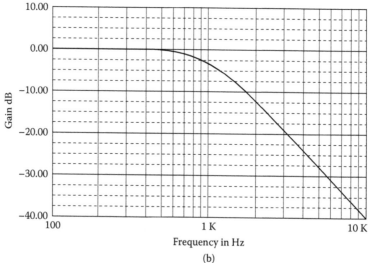

(b)

FIGURE 2.12 A 750-Hz LP Butterworth VCVS (a); its frequency response (b).

Hence,

$$C_1 = \frac{a + \sqrt{a^2 + 8b(K-1)}}{4b} = \frac{0.645 + \sqrt{0.645^2 + 8 \times 0.708 \times 4}}{4 \times 0.708} = 1.924 \ F$$

$$C_2 = \frac{1}{bC_1} = \frac{1}{0.708 \times 1.924} = \frac{1}{1.362} = 0.734 \ F$$

Denormalization

$$ISF = 10^4$$

$$FSF = \frac{\omega_1}{\omega_n} = \frac{2\pi f_1}{1} = 2\pi \times 300 = 600\pi$$

$$C_1 = \frac{C_{1n}}{ISF \times FSF} = \frac{1.924}{6\pi \times 10^6} = 102 \ nF$$

$$C_2 = \frac{C_{2n}}{ISF \times FSF} = \frac{0.734}{6\pi \times 10^6} = 38.9 \ nF$$

$$R = ISF \times R_n = 10^4 \times 1 \ \Omega = 10 \ k\Omega$$

$$R_b = (K-1)R_a \qquad \therefore \qquad R_a = 5 \ k\Omega \qquad \therefore \qquad R_b = 20 \ k\Omega$$

Figure 2.13a shows the designed filter, with its frequency response in Figure 2.13b.

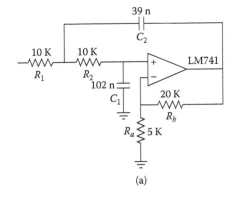

(a)

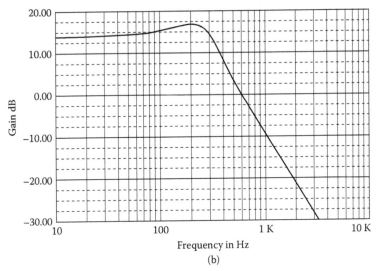

(b)

FIGURE 2.13 (a) Second-order LP Chebyshev 3-db VCVS, $f_1 = 300$ Hz, $K = 5$; (b) its frequency response.

EXAMPLE 2.6

Design a 1-kHz Bessel filter with gain 10.

Solution

From Bessel coefficients in Appendix C, we find:

$$a = 3.000, \ b = 3.000$$

Hence:

$$C_1 = \frac{a + \sqrt{a^2 + 8b(K-1)}}{4b} = \frac{3 + \sqrt{9 + 8 \times 3 \times 9}}{4 \times 3} = 1.5 \ \text{F}$$

$$C_2 = \frac{1}{bC_1} = \frac{1}{3 \times 1.475} = 0.222 \ \text{F}$$

Denormalization

$$ISF = 10^4$$

$$FSF = \frac{\omega_1}{\omega_n} = 2\pi f_1 = 2\pi \times 10^3$$

$$C_1 = \frac{C_{1n}}{ISF \times FSF} = \frac{1.475}{2\pi \times 10^7} = 23.8 \ \text{nF}$$

$$C_2 = \frac{C_{2n}}{ISF \times FSF} = \frac{0.226}{2\pi \times 10^7} = 3.5 \ \text{nF}$$

For $R_a = 2.2 \ \text{k}\Omega$ $\quad \therefore$

$$R_b = (K-1)R_a = 9 \times 2.2 \ \text{k}\Omega \cong 20 \ \text{k}\Omega$$

Figure 2.14a shows the designed filter, with its phase response in Figure 2.14b.

Second method (equal components)

In this method

$$C_{1n} = C_{2n} = C_n$$

From Equation (2.31), we have:

$$C_{1n} = C_{2n} = C_n = \frac{1}{\sqrt{b}} \tag{2.38}$$

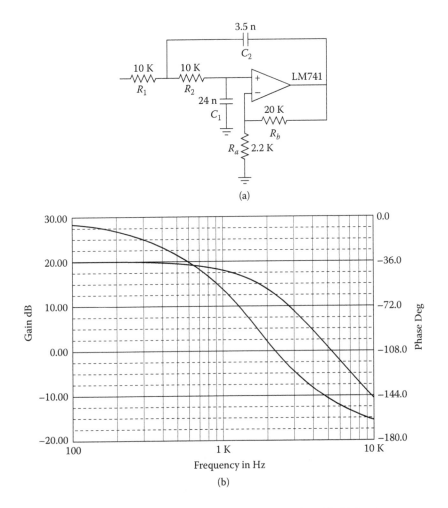

FIGURE 2.14 (a) Second-order LP Bessel filter, $f_1 = 1$ kHz, $K = 10$; (b) its phase response.

and Equation (2.29), becomes

$$\frac{3-K}{a} = a \qquad \therefore$$

$$K = 3 - \frac{a}{\sqrt{b}} \qquad\qquad (2.39)$$

We have an "equal component" VCVS low-pass filter, for a normalized cutoff frequency of $W_1 = 1$ rad/s. However, we pay a premium for the convenience of having equal resistors and capacitors. The passband gain will be fixed (Equation 2.39).

Example 2.7

Design a 1-kHz "equal-component" Butterworth low-pass filter.

Solution

$$C_n = 1 \text{ F}, \qquad R_n = 1 \, \Omega$$

From Butterworth coefficients of Appendix C, for $n = 2$, we find:

$$a = 1.414, \ b = 1.000$$

Hence:

$$K = 3 - \frac{a}{\sqrt{b}} = 3 - 1.414 = 1.586$$

Denormalization

$$ISF = 10^4$$

$$FSF = \frac{\omega_1}{\omega_n} = 2\pi f_1 = 2\pi \times 10^3$$

Hence:

$$R = ISF \times R_n = 10^4 \times 1 \, \Omega = 10 \text{ k}\Omega$$

$$R_a = ISF \times R_{an} = 10^4 \times 1 \, \Omega = 10 \text{ k}\Omega$$

$$R_b = (K - 1)R_{an} = 0.586 \times 10 \text{ k}\Omega = 5.9 \text{ k}\Omega$$

$$C = \frac{C_n}{ISF \times FSF} = \frac{1}{2\pi \times 10^7} = 15.9 \text{ nF}$$

Figure 2.15a shows the designed filter, with its frequency response in Figure 2.15b.

Example 2.8

Design a 1-kHz LP Chebyshev 1-dB "equal component" filter.

Solution

From Chebyshev 1-dB coefficients in Appendix C, we find:

$$a = 1.098 \text{ and } b = 1.103$$

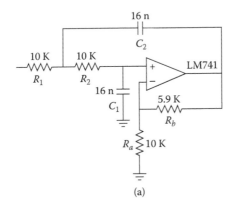

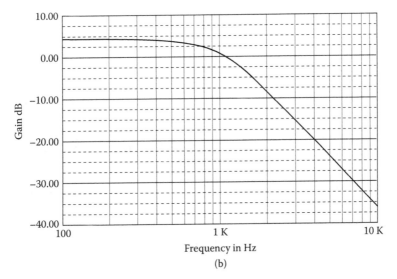

FIGURE 2.15 (a) Second-order LP Butterworth filter; (b) its frequency response.

Hence:

$$C_n = \frac{1}{\sqrt{b}} = \frac{1}{\sqrt{1.103}} = 0.950 \text{ F}$$

$$K = 3 - \frac{a}{\sqrt{b}} = 3 - \frac{1.098}{\sqrt{1.103}} = 1.954$$

Denormalization

$$ISF = 10^4$$

$$FSF = \frac{\omega_1}{\omega_n} = 2\pi f_1 = 2\pi \times 10^3$$

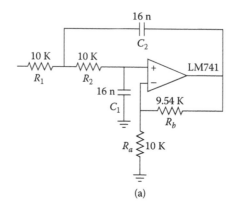

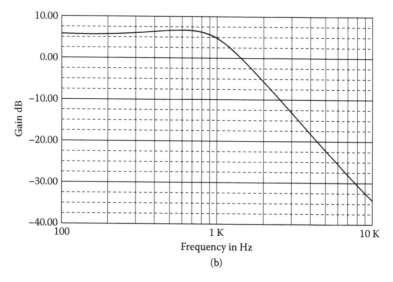

FIGURE 2.16 (a) Second-order LP Chebyshev 1-dB LPF; (b) its frequency response.

$$R = ISF \times R_n = 10^4 \times 1\,\Omega = 10\text{ k}\Omega$$

$$C = \frac{C_n}{ISF \times FSF} = \frac{1}{2\pi \times 10^7} = 15.9\text{ nF}$$

$$R_a = 10\text{ k}\Omega \qquad \therefore \qquad R_b = (K-1)R_{an} = 0.954 \times 10^4\,\Omega = 9.54\text{ k}\Omega$$

Figure 2.16a shows the designed filter, and its frequency response is shown in Figure 2.16b.

2.7 HIGH-PASS FILTERS

Active high-pass filters can be derived directly from the normalized low-pass configurations by a suitable transformation. To make the conversion, replace each resistor by a capacitor having the reciprocal value and vice versa as follows:

$$C_{HP} = \frac{1}{R_{LP}} \tag{2.40}$$

$$R_{HP} = \frac{1}{C_{LP}} \tag{2.41}$$

It is important to recognize that only the resistors that are part of the low-pass RC networks are transformed into capacitors by Equation (2.39). Feedback resistors that strictly determine operational amplifier gain, such as R_a and R_b, in Figure 2.9, are omitted from the transformation.

After the normalized low-pass configuration is transformed into a high-pass filter, the circuit is frequency- and impedance-scaled in the same manner as in the design of low-pass filters.

EXAMPLE 2.9

Design a 100-Hz HP Butterworth with gain 10.

Solution

From Butterworth coefficients in Appendix C, we have:

$$a = 1.414, \; b = 1.000$$

$$C_{1n} = \frac{a^2 + \sqrt{a^2 + 8b(K-1)}}{4b} = \frac{1.414 + \sqrt{1.414^2 + 8 \times 9}}{4} = 2.504 \text{ F} \qquad \therefore$$

$$C_{2n} = \frac{1}{bC_{1n}} = \frac{1}{2.504} = 0.399 \text{ F}$$

Hence:

$$R_{1n} = \frac{1}{C_{1n}} = \frac{1}{2.504} = 0.399 \; \Omega$$

$$R_{2n} = \frac{1}{C_{2n}} = \frac{1}{0.399} = 2.504 \; \Omega$$

$$C_{1n} = C_{2n} = C_n = 1 \text{ F}$$

Denormalization

$$ISF = 10^4$$

$$FSF = \frac{\omega_2}{\omega_n} = 2\pi f_2 = 2\pi \times 10^2$$

$$C_1 = C_2 = C = \frac{C_n}{ISF \times FSF} = \frac{1}{2\pi \times 10^6} = 159.2 \text{ nF}$$

$$R_1 = ISF \times R_{1n} = 10 \times 0.399 \text{ k}\Omega = 3.99 \text{ k}\Omega$$

$$R_2 = ISF \times R_{2n} = 10 \times 2.504 \text{ k}\Omega = 25.04 \text{ k}\Omega$$

$$R_a = 10 \text{ } \zeta\Omega \qquad \therefore \qquad R_b = (K-1)R_{an} = 9 \times 10 \text{ k}\Omega = 90 \text{ k}\Omega$$

Figure 2.17a shows the designed filter, with its frequency response in Figure 2.17b.

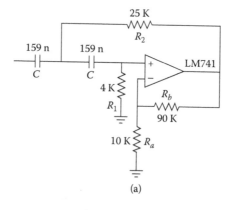

(a)

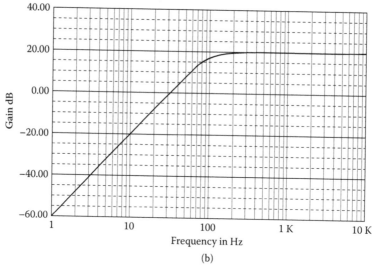

(b)

FIGURE 2.17 (a) HP Butterworth filter, with $f_2 = 100$ Hz, $K = 10$; (b) frequency response.

EXAMPLE **2.10**

A second-order HP Chebyshev 0.5-dB filter must be designed with a gain of 1 at a cutoff frequency of 100 Hz.

Solution

From Chebyshev 0.5-dB coefficients in Appendix C, we find:

$$a = 1.426 \text{ and } b = 1.516$$

$$C_{1n} = \frac{a}{2b} = \frac{1.426}{2 \times 1.516} = 0.470 \text{ F}$$

$$C_{2n} = \frac{2}{a} = \frac{2}{1.426} = 1.403 \text{ F}$$

Hence:

$$R_{1n} = \frac{1}{C_{1n}} = \frac{1}{0.470} = 2.126 \ \Omega$$

$$R_{2n} = \frac{1}{C_{2n}} = \frac{1}{1.403} = 0.713 \ \Omega$$

Denormalization

$$ISF = 10^4$$

$$FSF = \frac{\omega_2}{\omega_n} = 2\pi f_2 = 2\pi \times 10^2$$

$$C_1 = C_2 = C = \frac{C_n}{ISF \times FSF} = \frac{1}{2\pi \times 10^6} = 159.2 \text{ nF}$$

$$R_1 = ISF \times R_{1n} = 10 \times 2.126 \text{ k}\Omega = 21.3 \text{ k}\Omega$$

$$R_2 = ISF \times R_{2n} = 10 \times 0.713 \text{ k}\Omega = 7.1 \text{ k}\Omega$$

Figure 2.18a shows the designed filter, with its frequency response in Figure 2.18 b.

EXAMPLE **2.11**

Design an HP Chebyshev 3-dB "equal component" filter at a cutoff frequency of 200 Hz.

Solution

From Chebyshev 3-dB coefficients in Appendix C, we find:

$$a = 0.645, \ b = 0.708$$

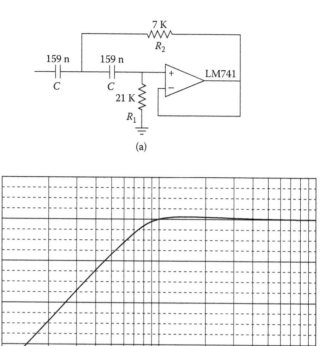

FIGURE 2.18 (a) HP Chebyshev 0.5 dB, $f_2 = 100$ Hz, $K = 1$; (b) its frequency response.

(a) For a low-pass filter:

$$R_{1n} = R_{2n} = R_n = 1\ \Omega$$

$$C_{1n} = C_{2n} = C_n = \frac{1}{\sqrt{b}} = \frac{1}{\sqrt{0.708}} = 1.188\ \text{F} \qquad \therefore$$

$$K = 3 - \frac{a}{\sqrt{b}} = 3 - \frac{0.645}{\sqrt{0.708}} = 3 - 0.767 = 2.233$$

(b) For a high-pass filter:

$$R_{1n} = R_{2n} = R_n = \frac{1}{C_n} = \frac{1}{1.188} = 0.841\ \text{F}$$

$$C_{1n} = C_{2n} = C_n = 1\ \text{F}$$

Denormalization

$$ISF = 10^4$$

$$FSF = \frac{\omega_2}{\omega_n} = 2\pi f_2 = 2\pi \times 200 = 4\pi \times 10^2$$

$$C_1 = C_2 = C = \frac{C_n}{ISF \times FSF} = \frac{1}{4\pi \times 10^6} = 79.6 \text{ nF}$$

$$R_1 = R_2 = R = ISF \times R_n = 10 \times 0.841 \ \Omega = 8.4 \text{ k}\Omega$$

$$R_a = 10 \text{ k}\Omega \quad \therefore \quad R_b = (K-1)R_a = (2.233-1) \times 10 \text{ k}\Omega$$

Figure 2.19a shows the designed filter, with its frequency response in Figure 2.19b.

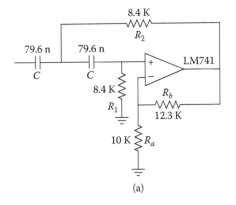

(a)

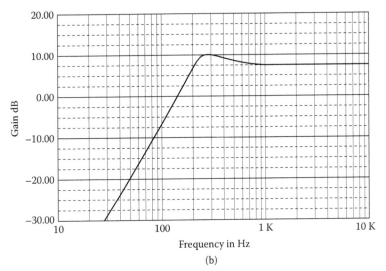

(b)

FIGURE 2.19 (a) HP Chebyshev 3-dB, "equal component," $f_2 = 200$ Hz; (b) its frequency response.

2.8 HIGHER-ORDER FILTERS

In the preceding sections of this chapter we have considered the realization of second-order filters using Sallen–Key circuits. Many filtering applications, however, require filters of higher than second order, either to provide greater stopband attenuation and sharper cutoff at the edge of the passband in the low-pass or high-pass case, or to provide a broad passband with some special transmission characteristic in the band-pass case. In this section we discuss some means of obtaining these higher-order filters.

One of the simplest approaches to the realization of higher-order filters is to factor the specified network function into quadratic factors, and realize each of these by a separate circuit using the second-order realizations given in preceding sections. Because these circuits all have an operational amplifier as an output element, their output impedance is low (in theory, zero) and, thus, a simple cascade of such second-order realizations may be made without interaction occurring between the individual stages. As a result, the overall voltage transfer function is simply the product of the individual transfer functions. An advantage of this approach is that each of the filters may be tuned separately, a point of considerable practical importance when high-order network functions are to be realized.

In the nth-order case, the general low-pass voltage transfer function is

$$H(s) = \frac{V_0(s)}{V_i(s)} = \frac{K}{s^n + b_{n-1}s^{n-1} + \ldots + b_1 s + b_0} \tag{2.42}$$

Such a function has a magnitude characteristic that decreases at the rate of $20n$ dB/dec or $6n$ dB/oct outside the passband. If n is even, Equation (2.42) may be written in the form

$$H(s) = \prod_{i=1}^{n/2} \frac{K_i}{s^2 + a_{i1}s + b_{i0}} \tag{2.43}$$

and the individual quadratic functions $K_i / (s^2 + a_{i1}s + a_{i0})$ may be synthesized by any of the second-order low-pass realizations given in the earlier sections. The realizations may be individually impedance-normalized to provide convenient element values. The constants K_i are, of course, arbitrary, unless we want to realize K exactly, in which case the product of the K_is must be equal K.

If n is odd, Equation (2.42) may be written as

$$H(s) = \frac{1}{s+b} \prod_{i=1}^{\frac{n-1}{2}} \frac{K_i}{s^2 + a_{i1}s + b_{i0}} \tag{2.44}$$

The quadratic factors are realized as before, and the first-order filter may be realized by the circuit shown in Figure 2.1 with unity gain.

Although it is theoretically possible to construct higher-order filters by cascading the necessary filter sections in any order we like, we will cascade the sections in the order of **decreasing damping**. This is the same as cascading the sections in the order of **increasing voltage gain**.

Occasionally, the filter specification may include all the necessary information, for example "design a fourth-order Butterworth low-pass filter with a 3-dB frequency equal to 200 Hz," but this would represent a rather trivial design exercise. It is much more likely that the designer will be presented with a requirement for a filter whose gain within the passband falls within a specified range, and whose gain in the stopband is less than or equal to some other specified value. If the frequency limits of the passband and the stopband are also given, the design can commence.

The following expression calculates the order n required for a Butterworth filter to meet a given set of attenuation/frequency specifications:

$$n = \frac{\log\left(\frac{10^{0.1A_{min}} - 1}{10^{0.1A_{max}} - 1}\right)}{2\log\left(\frac{f_s}{f_1}\right)} \tag{2.45}$$

When n has been determined, the attenuation in dB at f_s can be found from the expression:

$$A_{min} = 10\log\left[1 + (10^{0.1A_{max}} - 1)\left(\frac{f_s}{f_1}\right)^{2n}\right] \text{ dB} \tag{2.46}$$

Next, f_{3dB} must be found. Note that f_{3dB}, is the nominal cutoff frequency. f_{3dB} can be found from the expression:

$$f_{3dB} = f_1\left(\frac{10^{0.3} - 1}{10^{0.1A_{max}} - 1}\right)^{1/2n} \tag{2.47}$$

Consider a low-pass filter with $f_1 = 300$ Hz, $A_{max} = 1$ dB, $f_s = 500$ Hz, and $A_{min} = 20$ dB. Hence

$$n = \frac{\log\left(\frac{10^2 - 1}{10^{0.1} - 1}\right)}{2\log\left(\frac{500}{300}\right)} = 5.82$$

Because n must be an integer, we take the next highest value, which is 6.

$$A_{min} = 10\log\left[1 + (10^{0.1} - 1)\left(\frac{500}{300}\right)^{12}\right] \cong 20.8 \text{ dB}$$

$$f_{3dB} = 300 \left(\frac{10^{0.3} - 1}{10^{0.1} - 1} \right)^{1/12} = 300 \times 1.119 \ \text{Hz} = 335.7 \ \text{Hz}$$

The order n for a Chebyshev low-pass filter can be found from the expression:

$$n = \frac{\ln(x + \sqrt{x^2 - 1})}{\ln(y + \sqrt{y^2 - 1})} \tag{2.48}$$

where

$$x = \sqrt{\frac{10^{0.1A_{min}} - 1}{10^{0.1A_{max}} - 1}} \tag{2.49}$$

and

$$y = \frac{f_s}{f_1} \tag{2.50}$$

The attenuation in dB of a Chebyshev filter at any frequency f can be calculated from:

$$A(f) = 10\log\{1 + [\sqrt{10^{0.1A_{max}} - 1} \times 0.5(e^z + e^{-z})]^2\} \tag{2.51}$$

where

$$z = n\ln\left(\frac{f}{f_1} + \sqrt{\left(\frac{f}{f_1}\right)^2 - 1} \right) \tag{2.52}$$

Consider a low-pass Chebyshev filter with $f_1 = 1$ kHz, $f_s = 3$ kHz, $A_{max} = 0.2$ dB, and $A_{min} = 50$ dB. Hence

$$x = \sqrt{\frac{10^5 - 1}{10^{0.02} - 1}} = 1456.65$$

$$n = \frac{\ln(1456.65 + \sqrt{2121834})}{\ln(3 + \sqrt{8})} = 4.53 \quad \therefore \quad n = 5$$

$$z = 5\ln(3 + \sqrt{9 - 1}) = 8.8137$$

$$A(3\,\text{kHz}) = 10\log\{1 + [\sqrt{10^{0.02} - 1} \times 0.5(e^{8.814} + e^{-8.814})]^2\}$$

$$A(3\ \text{kHz}) = 57.3\ \text{dB}$$

Another way to determine n is to use the nomographs in Appendix B. They are used to determine the order n necessary to meet a given set of specifications. To use a nomograph, A_{max}, A_{min}, and the ratio f_s/f_1 or f_2/f_s (for HPF) must be known. Draw a straight line (1) from the desired value of A_{max} through A_{min} to the point of intersection with the left edge, corresponding to $f_s/f_1 = 1$, of the nomograph. Then draw a vertical line (2) corresponding to the desired value of f_s/f_1. Finally, draw a horizontal line (3) from the end of line (1) to line (2). The point of intersection of line (2) and line (3) should be between two of the numbered nomograph curves, and the required order n will be equal to the higher of the curves. Figure 2.20 illustrates the procedure.

EXAMPLE 2.12

Design an LP Butterworth filter to realize the following specifications: $f_1 = 3$ kHz, $f_s = 9$ kHz, $A_{max} = 3$ dB, $A_{min} = 40$ dB, and $K = 9$.

Solution

From the Butterworth nomographs we find the order of the filter, $n = 5$. From the Butterworth coefficients in Appendix C, we find:

First stage (first order)

$$b_0 = 1, \quad \text{hence:}$$

$$C_n = \frac{1}{b_0} = 1\,\text{F}, \qquad R_n = 1\ \Omega$$

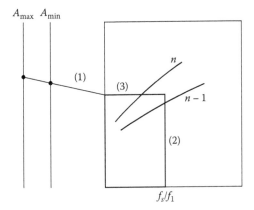

FIGURE 2.20 Example of nomograph use.

Second stage (second order)

$$a = 1.618, b = 1, K_1 = \sqrt{K} = \sqrt{9} = 3$$

hence:

$$C_{1n} = \frac{1.618 + \sqrt{1.618^2 + 8 \times 2}}{4} = 1.483 \text{ F}$$

$$C_{2n} = \frac{1}{1.483} = 0.674 \text{ F}$$

$$R_{1n} = R_{2n} = 1 \ \Omega$$

Third stage (third order)

$$a = 0.618, \ b = 1, \ K_2 = \sqrt{K} = \sqrt{9} = 3 \qquad \therefore$$

$$C_{1n} = \frac{0.618 + \sqrt{0.618^2 + 16}}{4} = 1.166 \text{ F}$$

$$C_{2n} = \frac{1}{1.166} = 0.857 \text{ F}, \qquad R_{1n} = R_{2n} = 1 \ \Omega$$

Denormalization

$$ISF = 10^4$$

$$FSF = \frac{\omega_1}{\omega_n} = 2\pi f_1 = 2\pi \times 3 \times 10^3 = 6\pi \times 10^3$$

First stage

$$R = ISF \times R_n = 10^4 \times 1 \ \Omega = 10 \ \text{k}\Omega$$

$$C = \frac{C_n}{ISF \times FSF} = \frac{1}{6\pi \times 10^7} = 5.31 \text{ nF}$$

Second stage

$$R_1 = R_2 = R = ISF \times R_n = 10^4 \times 1 \ \Omega = 10 \ \text{k}\Omega$$

$$C_1 = \frac{C_{1n}}{ISF \times FSF} = \frac{1.483}{6\pi \times 10^7} = 7.9 \text{ nF}$$

$$C_2 = \frac{C_{2n}}{ISF \times FSF} = \frac{0.674}{6\pi \times 10^7} = 3.6 \text{ nF}$$

$$R_a = 10 \text{ k}\Omega \quad \therefore \quad R_b = (K-1)R_a = 2 \times 10 \text{ k}\Omega = 20 \text{ k}\Omega$$

Third stage

$$R_1 = R_2 = R = ISF \times R_n = 10^4 \times 1\Omega = 10 \text{ k}\Omega$$

$$C_1 = \frac{C_{1n}}{ISF \times FSF} = \frac{1.166}{6\pi \times 10^7} = 6.2 \text{ nF}$$

$$C_2 = \frac{C_{2n}}{ISF \times FSF} = \frac{0.857}{6\pi \times 10^7} = 4.5 \text{ nF}$$

$$R_a = 10 \text{ k}\Omega, \qquad R_b = 20 \text{ k}\Omega$$

Figure 2.21a shows the designed filter, and its frequency response is shown in Figure 2.21b.

EXAMPLE 2.13

Design an HP Chebyshev 3-dB filter to realize the following specification: Gain 1, $f_2 = 1$ kHz, $f_s = 333$ Hz, and $A_{max} = 30$ dB.

Solution

From the Chebyshev nomographs in Appendix B, we find $n = 3$. From Chebyshev 3-dB coefficients in Appendix C, we have:
First stage (first order)

$$b = 0.299$$

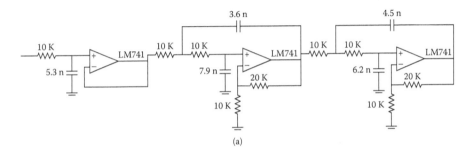

(a)

FIGURE 2.21 (a) LP Butterworth filter, $n = 5$, $f_1 = 3$ kHz, $K = 9$; (b) its frequency response.

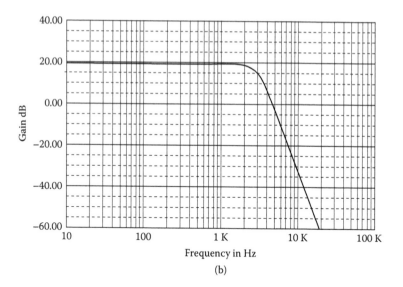

FIGURE 2.21 (Continued)

Low-pass filter:

$$R_n = 1 \ \Omega, \qquad C_n = \frac{1}{b} = \frac{1}{0.299} = 3.334 \ \text{F}$$

High-pass filter:

$$C_n = 1 \ \text{F}, \qquad R_n = \frac{1}{C_n} = \frac{1}{3.334} = 0.299 \ \Omega$$

Second stage (second order)

$$a = 0.299, \quad b = 0.839$$

Low-pass filter:

$$C_{1n} = \frac{0.299}{2 \times 0.839} = 0.178 \ \text{F}$$

$$C_{2n} = \frac{2}{0.299} = 6.689 \ \text{F}$$

$$R_n = 1 \ \Omega$$

High-pass filter:

$$R_{1n} = \frac{1}{C_{1n}} = \frac{1}{0.178} = 5.618 \ \Omega$$

$$R_{2n} = \frac{1}{C_{2n}} = \frac{1}{6.689} = 0.150 \ \Omega$$

$$C_{1n} = C_{2n} = C_n = 1 \ F$$

Denormalization

$$ISF = 10^4$$

$$FSF = \frac{\omega_2}{\omega_n} = 2\pi f_2 = 2\pi \times 10^3$$

First stage

$$R = ISF \times R_n = 10 \times 0.299 \ k\Omega = 2.99 \ k\Omega$$

$$C = \frac{C_n}{ISF \times FSF} = \frac{1}{2\pi \times 10^7} = 15.9 \ nF$$

Second stage

$$R_1 = ISF \times R_{1n} = 10 \times 5.621 \ \Omega = 56.2 \ k\Omega$$

$$R_2 = ISF \times R_{2n} = 10 \times 0.150 \ \Omega = 1.5 \ k\Omega$$

$$C = \frac{C_n}{ISF \times FSF} = \frac{1}{2\pi \times 10^7} = 15.9 \ nF$$

Figure 2.22a shows the designed filter, and its frequency response is shown in Figure 2.22b.

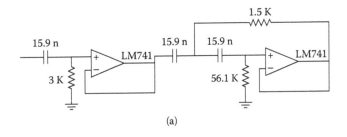

(a)

FIGURE 2.22 (a) HP Chebyshev 3-dB filter, $f_2 = 1$ kHz, $K = 1$; (b) its frequency response.

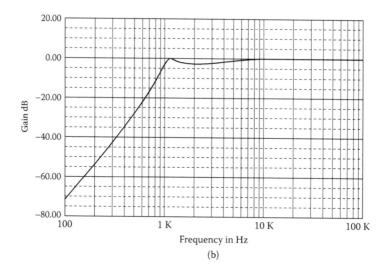

Frequency in Hz

(b)

FIGURE 2.22 (Continued)

EXAMPLE 2.14

Design an LP Bessel filter with the following characteristics: "equal component," third order, and cutoff frequency of 500 Hz.

Solution

From Bessel coefficients of Appendix C, we have:
First stage (first order)

$$b = 2.322 \qquad \therefore$$

$$C_n = \frac{1}{b} = \frac{1}{2.322} = 0.431 \text{ F}, \qquad R_n = 1 \text{ }\Omega$$

Second stage (second order)

$$a = 3.678, \quad b = 6.459$$

$$C_n = \frac{1}{\sqrt{b}} = \frac{1}{\sqrt{6.459}} = 0.393 \text{ }\Omega, \qquad R_n = 1 \text{ }\Omega$$

$$K = 3 - \frac{a}{\sqrt{b}} = 3 - \frac{3.678}{\sqrt{6.459}} = 1.553 \quad \text{or} \quad 3.8 \text{ dB}$$

Denormalization

$$ISF = 10^4$$

$$FSF = \frac{\omega_1}{\omega_n} = 2\pi f_1 = 2\pi \times 500 = \pi \times 10^3$$

First stage

$$R = ISF \times R_n = 10^4 \times 1 \; \Omega = 10 \; k\Omega$$

$$C = \frac{C_n}{ISF \times FSF} = \frac{0.431}{\pi \times 10^7} = 13.7 \; nF$$

Second stage

$$R = ISF \times R_n = 10^4 \times 1 \; \Omega = 10 \; k\Omega$$

$$C = \frac{C_n}{ISF \times FSF} = \frac{0.393}{\pi \times 10^7} = 12.5 \; nF$$

$$R_a = 10 \; k\Omega \qquad \therefore \qquad R_b = (K-1)R_a = 0.553 \times 10 \; k\Omega = 5.53 \; k\Omega$$

Figure 2.23a shows the designed filter, and its frequency response is shown in Figure 2.23b.

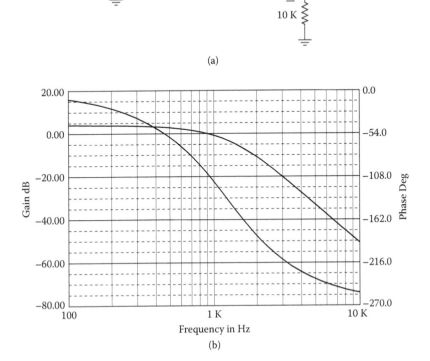

(a)

(b)

FIGURE 2.23 (a) Bessel filter, "equal component," $f_1 = 500$ Hz; (b) frequency and phase response.

2.9 WIDE-BAND FILTERS

Very often, particularly in audio applications, it is desired to pass a **wide band of frequencies** with relatively constant gain, as illustrated in Figure 2.24. Such a band-pass response is said to be characteristic of a **wide-band** filter.

When the separation between the upper and the lower cutoff frequencies (f_1, f_2) exceeds a ratio of approximately 2, the band-pass filter is considered a wide-band filter. The specifications are then separated into individual low-pass and high-pass requirements and met by a cascade of active low-pass filters having a cutoff frequency of f_1 and a high-pass filter having a cutoff frequency of f_2.

EXAMPLE 2.15

Design a 100–1000 Hz, Butterworth wide-band filter with the following specifications: $A_{max} = 3$ dB, $A_{min} = 30$ dB at $f_s = 40$ Hz, and 2500 Hz, with $K = 9$.

Solution

(a) *High-pass filter*
For $A_{max} = 3$ dB, $A_{min} = 30$ dB, $f_2 = 100$ Hz, and $f_{s2} = 40$ Hz, we find $n = 4$ ($f_2/f_{s2} = 100/40 = 2.5$).

From the Butterworth coefficients in Appendix C, we find:
First stage

$$a = 1.848, \quad b = 1.000$$

$$K_{HP} = \sqrt{K} = \sqrt{9} = 3 \quad \therefore \quad K_1 = K_2 = \sqrt{3} = 1.732$$

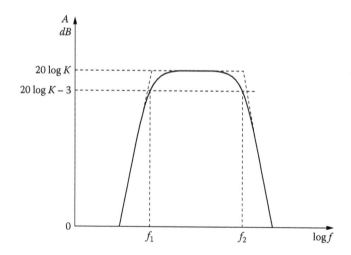

FIGURE 2.24 Characteristic band-pass response of a wide-band filter.

$$C_{1n} = \frac{1.848 + \sqrt{1.848^2 + 8 \times 0.732}}{4} = 1.223 \text{ F}$$

$$C_{2n} = \frac{1}{1.223} = 0.818 \text{ F} \qquad \therefore$$

$$R_{1n} = \frac{1}{C_{1n}} = \frac{1}{1.223} = 0.818 \text{ } \Omega$$

$$R_{2n} = \frac{1}{C_{2n}} = \frac{1}{0.818} = 1.223 \text{ } \Omega$$

Second stage (second order)

$$a = 0.765, \quad b = 1.000$$

$$C_{1n} = \frac{0.765 + \sqrt{0.765^2 + 8 \times 0.732}}{4} = 0.826 \text{ F}$$

$$C_{2n} = \frac{1}{0.826} = 1.211 \text{ F} \qquad \therefore$$

$$R_{1n} = \frac{1}{C_{1n}} = \frac{1}{0.826} = 1.211 \text{ } \Omega$$

$$R_{2n} = \frac{1}{C_{2n}} = \frac{1}{1.211} = 0.826 \text{ } \Omega$$

Denormalization

$$ISF = 10^4$$

$$FSF = \frac{\omega_2}{\omega_n} = 2\pi f_2 = 2\pi \times 100$$

First stage

$$R_1 = ISF \times R_{1n} = 10^4 \times 0.818 \text{ } \Omega \cong 8.2 \text{ k}\Omega$$

$$R_2 = ISF \times R_{2n} = 10^4 \times 1.223 \text{ } \Omega \cong 12 \text{ k}\Omega$$

$$C_1 = C_2 = C = \frac{C_n}{ISF \times FSF} = \frac{1}{2\pi \times 10^6} \cong 159 \text{ nF}$$

$$R_a = 10 \text{ k}\Omega \qquad \therefore \qquad R_b = (K-1)R_a = 0.732 \times 10 \text{ k}\Omega \cong 7.3 \text{ k}\Omega$$

Second stage

$$R_1 = ISF \times R_{1n} = 10^4 \times 1.211 \ \Omega \cong 12 \ k\Omega$$

$$R_2 = ISF \times R_{2n} = 10^4 \times 0.826 \ \Omega \cong 8.2 \ k\Omega$$

$$C = \frac{C_n}{ISF \times FSF} = \frac{1}{2\pi \times 10^6} \cong 159 \ nF$$

$$R_a = 10 \ k\Omega \quad \therefore \quad R_b \cong 7.3 \ k\Omega$$

(b) *Low-pass filter*

For $A_{max} = 3$ dB, $A_{min} = 30$ dB, and $f_{s1}/f_1 = 2500/1000 = 2.5$, we find $n = 4$.

From the Butterworth coefficients in Appendix C, we find:

First stage

$$a = 1.848, \quad b = 1.000$$

$$K_{HP} = \sqrt{K} = \sqrt{9} \quad \therefore \quad K_1 = K_2 = \sqrt{3} = 1.732$$

$$C_{1n} = \frac{1.848 + \sqrt{1.848^2 + 8 \times 0.732}}{4} = 1.223 \ F$$

$$C_{2n} = \frac{1}{C_{1n}} = \frac{1}{1.223} = 0.818 \ F$$

$$R_{1n} = R_{2n} = R_n = 1 \ \Omega$$

Second stage

$$a = 0.765, \qquad b = 1.000$$

$$C_{1n} = \frac{0.765 + \sqrt{0.765^2 + 8 \times 0.732}}{4} = 0.826 \ F$$

$$C_{2n} = \frac{1}{0.826} = 1.211 \ F$$

$$R_{1n} = R_{2n} = R_n = 1 \ \Omega$$

Denormalization

$$ISF = 10^4, \quad FSF = \frac{\omega_1}{\omega_n} = 2\pi f_1 = 2\pi \times 10^3, \quad ISF \times FSF = 2\pi \times 10^7$$

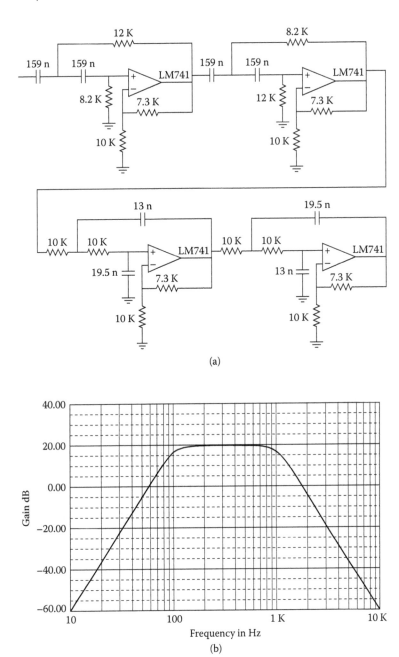

FIGURE 2.25 (a) A 100–1000 Hz wide-band filter; (b) frequency response.

First stage

$$R_1 = R_2 = R = ISF \times R_n = 10^4 \times 1 \ \Omega = 10 \ \text{k}\Omega$$

$$C_1 = \frac{C_{1n}}{ISF \times FSF} = \frac{1.223}{2\pi \times 10^7} = 19.5 \ \text{nF}$$

$$C_2 = \frac{C_{2n}}{ISF \times FSF} = \frac{0.818}{2\pi \times 10^7} = 13 \ \text{nF}$$

$$R_a = 10 \ \text{k}\Omega \qquad \therefore \qquad R_b = (K-1)R_a = 7.3 \ \text{k}\Omega$$

Second stage

$$R_1 = R_2 = R = 10^4 \times 1 \Omega = 10 \ \text{k}\Omega$$

$$C_1 = \frac{C_{1n}}{ISF \times FSF} = \frac{0.826}{2\pi \times 10^7} \cong 13 \ \text{nF}$$

$$C_2 = \frac{C_{2n}}{ISF \times FSF} = \frac{1.211}{2\pi \times 10^7} \cong 19.5 \ \text{nF}$$

$$R_a = 10 \ \text{k}\Omega \qquad \therefore \qquad R_b = 7.3 \ \text{k}\Omega$$

Figure 2.25a shows the designed filter, and its frequency response is shown in Figure 2.25b.

2.10 WIDE-BAND BAND-REJECT FILTERS

Wide-band band-reject filters can be designed by first separating the specification into individual low-pass and high-pass requirements. Low-pass and high-pass filters are then independently designed and combined by paralleling the inputs and summing both outputs to form the band-reject filter.

A wide-band approach is valid when the separation between cutoffs is an octave or more so that minimum interaction occurs in the stopband when the outputs are summed.

An inverting amplifier is used for summing and can also provide gain. Filters can be combined using the configuration of Figure 2.26, where R is arbitrary and K is the desired gain. The individual filters should have a low output impedance to avoid loading by the summing resistors.

Example 2.16

Design a band-reject Butterworth filter having 3-dB points at 100 and 400 Hz and greater than 30 dB of attenuation between 180 and 222 Hz with gain 1.

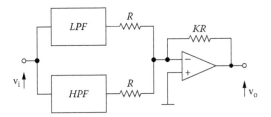

FIGURE 2.26 Block diagram of a wide-band band-reject filter.

Solution

Because the ratio of upper cutoff to lower is well in excess of an octave, a wide-band approach can be used. First, separate the specification into individual low-pass and high-pass requirements.

(a) *Low-pass filter*

$$A_{max} = 3 \text{ dB}, \quad A_{min} = 30 \text{ dB}, \quad f_{s1} / f_1 = 180 / 100 = 1.8$$

and from the Butterworth nomographs, we find $n = 4$.

First stage

From the Butterworth coefficients, for $n = 4$, we find:

$$a = 1.848, \quad b = 1.000 \quad \therefore$$

$$C_{1n} = \frac{1.848}{2} = 0.924 \text{ F}$$

$$C_{2n} = \frac{2}{1.848} = 1.082 \text{ F}$$

$$R_n = 1 \ \Omega$$

Second stage

$$a = 0.765, \quad b = 1.000$$

$$C_{1n} = \frac{0.765}{2} = 0.383 \text{ F}$$

$$C_{2n} = \frac{2}{0.765} = 2.614 \text{ F}$$

$$R_n = 1 \ \Omega$$

Denormalization

$$ISF = 10^4, \quad FSF = \frac{\omega_1}{\omega_n} = 2\pi f_1 = 2\pi \times 10^2$$

First stage

$$R = ISF \times R_n = 10^4 \times 1 \ \Omega = 10 \ \text{k}\Omega$$

$$C_1 = \frac{C_{1n}}{ISF \times FSF} = \frac{0.924}{2\pi \times 10^6} \cong 147 \ \text{nF}$$

$$C_2 = \frac{C_{2n}}{ISF \times FSF} = \frac{1.082}{2\pi \times 10^6} \cong 172 \ \text{nF}$$

Second stage

$$R = ISF \times R_n = 10^4 \times 1 \ \Omega = 10 \ \text{k}\Omega$$

$$C_1 = \frac{C_{1n}}{ISF \times FSF} = \frac{0.383}{2\pi \times 10^6} \cong 61 \ \text{nF}$$

$$C_2 = \frac{C_{2n}}{ISF \times FSF} = \frac{2.614}{2\pi \times 10^6} = 416 \ \text{nF}$$

(b) *High-pass filter*
For $A_{max} = 3$ dB, $A_{min} = 30$ dB, $f_2 / f_{s2} = 400 / 222 \cong 1.8$, and from the Butterworth nomographs, we find $n = 4$.
First stage

$$a = 1.848, \quad b = 1.000$$

$$C_{1n} = \frac{1.848}{2} = 0.924 \ \text{F}$$

$$C_{2n} = \frac{2}{1.848} = 1.082 \ \text{F} \quad \therefore$$

$$R_{1n} = \frac{1}{C_{1n}} = \frac{1}{0.924} = 1.082 \ \Omega$$

$$R_{2n} = \frac{1}{C_{2n}} = \frac{1}{1.082} = 0.924 \ \Omega$$

$$C_n = 1 \ \text{F}$$

Second stage

$$a = 0.765, \quad b = 1.000$$

$$C_{1n} = \frac{0.765}{2} = 0.378 \ \text{F}$$

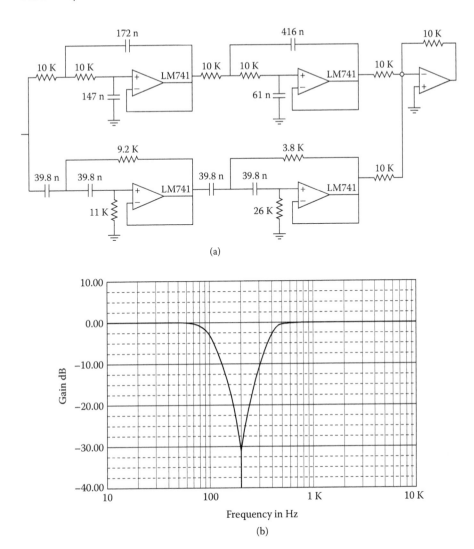

FIGURE 2.27 (a) Wide-band band-reject filter; (b) its frequency response.

$$C_{2n} = \frac{2}{0.765} = 2.614 \text{ F} \qquad \therefore$$

$$R_{1n} = \frac{1}{C_{1n}} = \frac{1}{0.383} = 2.614 \ \Omega$$

$$R_{2n} = \frac{1}{C_{2n}} = \frac{1}{2.614} = 0.383 \ \Omega$$

$$C_n = 1 \text{ F}$$

Denormalization

$$ISF = 10^4, \quad FSF = \frac{\omega_2}{\omega_n} = 2\pi f_2 = 2\pi \times 400 = 8\pi \times 10^2$$

$$C = \frac{C_n}{ISF \times FSF} = \frac{1}{8\pi \times 10^6} = 39.8 \ nF$$

First stage

$$R_1 = ISF \times R_{1n} = 10^4 \times 1.088 \ \Omega = 10.9 \ k\Omega$$

$$R_2 = ISF \times R_{2n} = 10^4 \times 0.924 \ \Omega = 9.2 \ k\Omega$$

Second stage

$$R_1 = ISF \times R_{1n} = 10^4 \times 2.614 \ \Omega \cong 26.1 \ k\Omega$$

$$R_2 = ISF \times R_{2n} = 10^4 \times 0.383 \ \Omega \cong 3.8 \ k\Omega$$

Figure 2.27a shows the designed wide-band band-reject filter, and its frequency response is shown in Figure 2.27b.

2.11 COMMENTS ON VCVS FILTERS

2.11.1 LOW-PASS FILTERS

1. In the case of multiple-stages ($n > 2$), the K parameter for each stage need not to be the same.
2. For best performance, the input resistance of the op-amp should be at least 10 times $R_{eq} = R + R = 2R$.
3. Standard resistance values of 5% tolerance normally yield acceptable results in the lower-order cases. For fifth and sixth orders, resistances of 2% tolerance probably should be used, and for seventh and eighth orders, 1% tolerances probably should be used.
4. In the case of capacitors, percentage tolerances should parallel those given earlier for the resistors for best results. Because precision capacitors are relatively expensive, it may be desirable to use capacitors of higher tolerances, in which case trimming is generally required. For lower orders ($n \leq 4$), 10% tolerance capacitors are quite often satisfactory.
5. The gain of each stage of the filter is $1 + R_b/R_a$, which can be adjusted to the correct value by using a potentiometer in lieu of resistors R_a and R_b. This is accomplished by connecting the center tap of the potentiometer to the inverting input of the op-amp. These gain adjustments are very useful in tuning the overall response of the filter.

6. In the low-pass filter section, maximum gain peaking is very nearly equal to Q at f_1. So, as a rule of thumb:
   (a) The op-amp bandwidth (BW) should be at least

$$BW = 100 \times K \times Q^3 \times f_1 \qquad (Q = 1/a)$$

For the real-pole section:

$$BW = 50 \times f_1$$

(b) For adequate full-power response, the slew rate (SR) of the op-amp must be

$$SR > \pi \times V_{Opp} \times BW_f$$

where BW_f is the filter bandwidth.

EXAMPLE 2.17

A unity gain 20-kHz 5-pole Butterworth filter would require:

$$BW_1 = 100 \times 1 \times 0.618^3 \times 20 \times 10^3 = 472 \ \text{kHz}$$

$$BW_2 = 100 \times 1 \times 1.618^3 \times 20 \times 10^3 = 8.77 \ \text{MHz}$$

$$BW_3 = 50 \times 20 \times 10^3 = 1 \ \text{MHz}$$

The worst case is 8.8-MHz op-amp BW.

$$SR > \pi \times 20 \times 20 \times 10^3 \ V/s > 1.3 \ V/\mu s \quad (\text{for 20-V p-p output})$$

2.11.2 HIGH-PASS FILTERS

1. For multiple-stage filters ($n > 2$), the K parameter for each stage need not be the same.
2. For best performance, the input resistance of the op-amp should be at least 10 times $R_{eq} = R_1$.
3. Standard resistance values of 5% tolerance normally yield acceptable results in the lower-order cases. For fifth and sixth orders, resistances of 2% tolerance probably should be used, and for seventh and eighth orders, 1% tolerance probably should be used.
4. In the case of capacitors, percentage tolerances should parallel those given earlier for the resistors, for best results. Because precision capacitors are relatively expensive, it may be desirable to use capacitors of higher tolerances,

in which case trimming is generally required. In the case of the lower orders ($n \leq 4$), 10% tolerance capacitors are quite often satisfactory.

5. The gain of each stage of the filters is $1 + R_b/R_a$, which can be adjusted to the corrected value by using a potentiometer in lieu of resistors R_a and R_b. This is accomplished by connecting the center tap of the potentiometer to the inverting input of the op-amp. These gain adjustments are very useful in tuning the overall response of the filter.

PROBLEMS

2.1 Design an "equal component" Butterworth band-pass filter having 3 dB at 300 and 3000 Hz and attenuation greater than 30 dB between 50 and 18,000 Hz.

2.2 Design an "equal component" Chebyshev 2-dB band-reject filter at 250 and 1500 Hz and attenuation greater than 30 dB between 500 and 750 Hz.

2.3 For the high-pass filter:
 a. Find the transfer function.
 b. Find the formulas for the normalized filter, i.e., $C = 1$ F and $w_2 = 1$ rad/s.

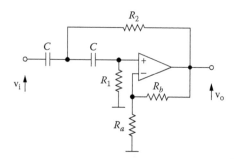

2.4 For the following narrow-band band-reject filter:

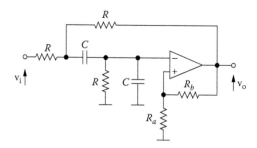

 a. Find the transfer function.
 b. Find the formulas of the normalized filter.
 c. Design the filter for $f_o = 100$ Hz and Q ($1/a$) = 5.

2.5 Design a Chebyshev 3-dB LPF with $f_1 = 3000$ Hz and attenuation 40 dB at 10.5 kHz with gain 1.

2.6 Design a Chebyshev 3-dB HPF with $f_2 = 300$ Hz and attenuation 40 dB at 85.7 Hz with gain 1.

2.7 Design an "equal component" Butterworth LPF with 3 dB at 10 kHz and 50 dB at 35 kHz.

2.8 Design a Butterworth "equal component" HPF with 3 dB at 1 kHz and 50 dB at 285.7 kHz.

2.9 Design a Bessel filter with 3 dB at 1 kHz and 45 dB at 4 kHz, with unity gain.

2.10 For the following filter:

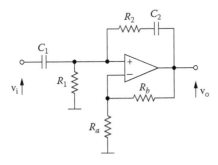

a. Find the transfer function.

b. For $R_1 = R_2 = R = 1\ \Omega$, $C_1 = C_2 = C = 1$ F, find the formulas to design the filter.

3 MultiFeedback Filters

In this chapter we present a different approach to the design of active filters, namely, the use of the entire passive RC network to provide the feedback around the operational amplifier. A general configuration for the second-order case is shown in Figure 3.1. It is called a **multifeedback infinite gain amplifier filter**. The resulting transfer function is an inverting one, i.e., the dc gain is negative. With this configuration we can have a low-pass, a high-pass, and narrow-band band-pass characteristic.

3.1 LOW-PASS FILTERS

The basic circuit of the second-order infinite gain MFB (multifeedback) low-pass filter is shown in Figure 3.2. By using nodal analysis the transfer function can now be solved. For an ideal op-amp we have:

node v_1

$$-GV_i + (G_1 + G_2 + G_3 + sC_1)V_1 - G_3V_2 - G_2V_0 = 0 \qquad (3.1)$$

node v_2

$$-G_3V_1 + (G_3 + sC_2)V_2 - sC_2V_0 = 0 \qquad (3.2)$$

$$V_2 \cong 0 \qquad (3.3)$$

From Equations (3.3) and (3.2), we have:

$$V_1 = -\frac{sC_2}{G_3}V_0 \qquad (3.4)$$

From Equations (3.1), (3.3), and (3.4), we have:

$$H(s) = \frac{V_0}{V_i} = -\frac{G_1G_3}{s^2C_1C_2 + sC_2(G_1 + G_2 + G_3) + G_2G_3} \qquad \therefore$$

$$H(s) = -\frac{\dfrac{G_1G_3}{C_1C_2}}{s^2 + \dfrac{G_1 + G_2 + G_3}{C_1}s + \dfrac{G_2G_3}{C_1C_2}} \qquad (3.5)$$

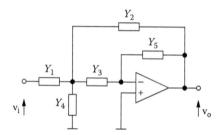

FIGURE 3.1 General multifeedback infinite gain amplifier filter.

For $s = 0$, we have:

$$K = H(0) = -\frac{G_1}{G_2} = -\frac{R_2}{R_1} \tag{3.6}$$

Hence, Equation (3.5) can be written as

$$H(s) = -\frac{K}{s^2 + as + b} \tag{3.7}$$

where

$$a = \frac{G_1 + G_2 + G_3}{C_1} \tag{3.8}$$

and

$$b = \omega_1^2 = \frac{G_2 G_3}{C_1 C_2} \tag{3.9}$$

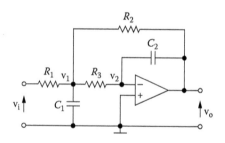

FIGURE 3.2 An infinite gain multifeedback low-pass filter.

If we set

$$R_1 = R_3 = R = 1 \ \Omega \qquad (3.10)$$

from Equation (3.6), we have:

$$R_2 = K \qquad (3.11)$$

and from Equations (3.8) and (3.9), we have, respectively:

$$\frac{2+G_2}{C_1} = a \qquad (3.12)$$

$$\frac{G_2}{C_1 C_2} = b \qquad (3.13)$$

From Equations (3.11) and (3.6), we have:

$$\frac{2+\dfrac{1}{KR}}{C_1} = a \qquad \therefore$$

$$C_1 = \frac{2K+1}{aK} \qquad (3.14)$$

From Equations (3.11), (3.13), and (3.14), we have:

$$C_2 = \frac{G_2}{bC_1} = \frac{1}{bR_2 C_1} = \frac{1}{bKC_1} \qquad \therefore$$

$$C_2 = \frac{a}{(2K+1)b} \qquad (3.15)$$

EXAMPLE 3.1

Design a second-order LP Butterworth filter with a gain of 10 at a cutoff frequency of 1 kHz.

Solution

From the Butterworth coefficients in Appendix C, for $n = 2$, we find:

$$a = 1.414 \qquad b = 1.000$$

Hence,

$$C_{1n} = \frac{2K+1}{aK} = \frac{21}{14.14} = 1.485 \ \text{F}$$

$$C_{2n} = \frac{a}{(2K+1)b} = \frac{1.414}{21} = 0.067 \ \text{F}$$

Denormalization

$$ISF = 10^4, \quad FSF = \frac{\omega_1}{\omega_n} = 2\pi f_1 = 2\pi \times 10^3$$

$$C_1 = \frac{C_{1n}}{ISF \times FSF} = \frac{1.485}{2\pi \times 10^7} = 23.6 \ \text{nF}$$

$$C_2 = \frac{C_{2n}}{ISF \times FSF} = \frac{0.067}{2\pi \times 10^7} = 1.1 \ \text{nF}$$

$$R = ISF \times R_n = 10^4 \times 1 \ \Omega = 10 \ \text{k}\Omega$$

$$R_2 = KR = 10 \times 10^4 \ \Omega = 100 \ \text{k}\Omega$$

Figure 3.3a shows the designed filter, and its frequency response is shown in Figure 3.3b.

EXAMPLE 3.2

Design a second-order LP Chebyshev 3-dB filter with a gain of 1 and cutoff frequency of 1 kHz.

Solution

From Chebyshev 3-dB coefficients in Appendix C, for $n = 2$, we find:

$$a = 0.645 \quad b = 0.708$$

Hence,

$$R_n = 1 \ \Omega, R_{2n} = KR_n = 1 \times 1 \ \Omega = 1 \ \Omega$$

$$C_{1n} = \frac{2K+1}{aK} = \frac{3}{0.645} = 4.651 \, \text{F}$$

$$C_{2n} = \frac{a}{(2K+1)b} = \frac{0.645}{3 \times 0.708} = 0.304 \ \text{F}$$

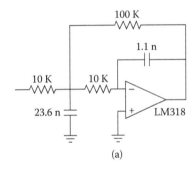

(a)

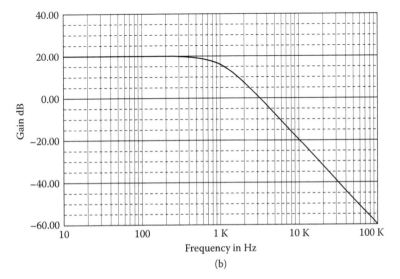

(b)

FIGURE 3.3 (a) Second-order LP Butterworth filter, $f_1 = 1$ kHz, $K = 10$; (b) its frequency response.

Denormalization

$$ISF = 10^4$$

$$FSF = \frac{\omega_1}{\omega_n} = \frac{2\pi f_1}{1} = 2\pi \times 10^3$$

$$R = ISF \times R_n = 10^4 \times 1 \ \Omega = 10 \ k\Omega$$

$$R_2 = ISF \times R_{2n} = 10^4 \times 1 \ \Omega = 10 \ k\Omega$$

$$C_1 = \frac{C_{1n}}{ISF \times FSF} = \frac{4.651}{2\pi \times 10^7} = 74 \ nF$$

$$C_2 = \frac{C_{2n}}{ISF \times FSF} = \frac{0.304}{2\pi \times 10^7} = 4.8 \ nF$$

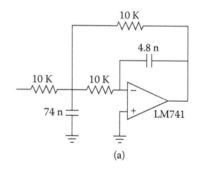

(a)

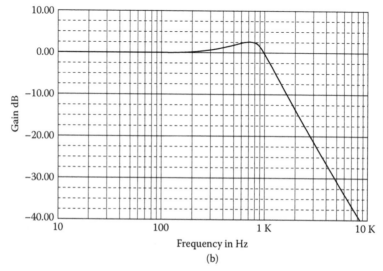

Frequency in Hz

(b)

FIGURE 3.4 (a) Second-order LP Chebyshev 3 dB filter, $f_1 = 1$ kHz, $K = 1$; (b) frequency response; (c) its ripple.

Figure 3.4a shows the designed filter, with its frequency response in Figure 3.4b [From mCap III].

3.2 HIGH-PASS FILTERS

High-pass filters can be derived directly from the normalized low-pass configurations by suitable transformation, as in the case of Sallen–Key filters.

Figure 3.5 shows the high-pass MFB filter.

From Figure 3.5b, for the normalized HPF, we have:

$$C_1 = C_3 = C = 1 \text{ F} \tag{3.16}$$

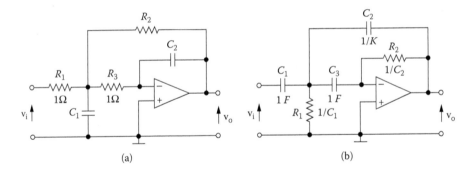

FIGURE 3.5 LP- to HP-transformation.

$$K = \frac{\dfrac{1}{\omega C_2}}{\dfrac{1}{\omega C_1}} = \frac{C_1}{C_2} \qquad \therefore$$

$$C_2 = \frac{C_1}{K} = \frac{1}{K} \tag{3.17}$$

EXAMPLE **3.3**

Design a second-order HP Butterworth filter with a gain of 5 and a cutoff frequency of 100 Hz.

Solution

(a) *Low-pass filter*
From Butterworth coefficients of Appendix C, for $n = 2$, we have:

$$a = 1.414 \qquad b = 1.000$$

Hence:

$$C_{1n} = \frac{2K + 1}{aK} = \frac{11}{1.414 \times 5} = 1.556 \ \text{F}$$

$$C_{2n} = \frac{a}{(2K + 1)b} = \frac{1.414}{11 \times 1} = 0.129 \ \text{F}$$

(b) *High-pass filter*

$$R_{1n} = \frac{1}{C_{1n}} = \frac{1}{1.556} = 0.647 \ \Omega$$

$$R_{2n} = \frac{1}{C_{2n}} = \frac{1}{0.129} = 7.752 \ \Omega$$

$$C_{1n} = C_{3n} = C_n = 1 \ \text{F}$$

$$C_{2n} = \frac{C_n}{K} = \frac{1}{5} = 0.2 \ \text{F}$$

Denormalization

$$ISF = 10^4$$

$$FSF = \frac{\omega_2}{\omega_n} = 2\pi f_2 = 2\pi \times 10^2 \qquad \therefore$$

$$R_1 = ISF \times R_{1n} = 10^4 \times 0.643 \ \Omega = 6.4 \ \text{k}\Omega$$

$$R_2 = ISF \times R_{2n} = 10^4 \times 7.752 \ \Omega = 77.5 \ \text{k}\Omega$$

$$C_1 = C_3 = \frac{C_n}{ISF \times FSF} = \frac{1}{2\pi \times 10^6} = 159.2 \ \text{nF}$$

$$C_2 = \frac{C_{2n}}{ISF \times FSF} = \frac{0.2}{2\pi \times 10^6} = 31.8 \ \text{nF}$$

Figure 3.6a shows the designed filter, and its frequency response is shown in Figure 3.6b.

EXAMPLE 3.4

Design a second-order HP Chebyshev 3-dB filter with a gain of 10 and a cutoff frequency of 100 Hz.

Solution

From Chebyshev 3-dB coefficients of Appendix C, for $n = 2$, we have:

$$a = 0.645 \qquad b = 0.708$$

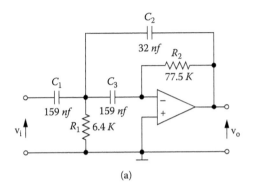

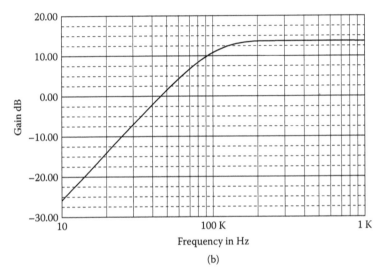

FIGURE 3.6 (a) Second-order HP Butterworth filter, $f_2 = 100$ Hz, $K = 5$; (b) frequency response.

(a) *Low-pass filter*

$$C_{1n} = \frac{2K+1}{aK} = \frac{21}{0.645 \times 10} = 3.256 \text{ F}$$

$$C_{2n} = \frac{a}{(2K+1)b} = \frac{0.645}{21 \times 0.708} = 0.043 \text{ F}$$

(b) *High-pass filter*

$$R_{1n} = \frac{1}{C_{1n}} = \frac{1}{3.256} = 0.307 \ \Omega$$

$$R_{2n} = \frac{1}{C_{2n}} = \frac{1}{0.043} = 23.051 \ \Omega$$

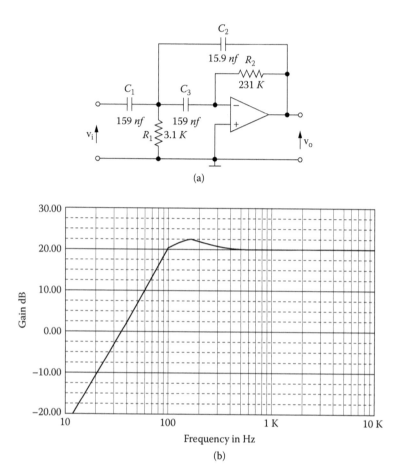

FIGURE 3.7 (a) Second-order HP Chebyshev 3 dB, $f_2 = 100$ Hz, $K = 10$; (b) frequency response.

$$C_{1n} = C_{3n} = C_n = 1 \text{ F}$$

$$C_{2n} = \frac{C_{2n}}{K} = \frac{1}{10} = 0.1 \text{ F}$$

Denormalization

$$ISF = 10^4, \quad FSF = \frac{\omega_2}{\omega_n} = 2\pi f_2 = 2\pi \times 10^2 \quad \therefore$$

$$C_1 = C_3 = C = \frac{C_n}{ISF \times FSF} = \frac{1}{2\pi \times 10^6} = 159.2 \text{ nF}$$

$$C_2 = \frac{C_{2n}}{ISF \times FSF} = \frac{0.1}{2\pi \times 10^6} = 15.9 \text{ nF}$$

$$R_1 = ISF \times R_{1n} = 10^4 \times 0.307 \ \Omega \cong 3.1 \text{ k}\Omega$$

$$R_2 = ISF \times R_{2n} = 10^4 \times 23.051 \ \Omega \cong 231 \text{ k}\Omega$$

Figure 3.7a shows the designed filter, and its frequency response is shown in Figure 3.7b.

3.3 HIGHER-ORDER FILTERS

We may obtain higher-order low-pass Butterworth, Chebyshev, or Bessel filters and high-pass Butterworth or Chebyshev filters by cascading two or more networks until the order of filter that the designer desires is attained. With these configurations we can have only (usually) even-order filters.

EXAMPLE 3.5

Design an LP Butterworth filter 3 dB at 1 kHz and attenuation 35 dB at 4 kHz with a gain of 5.

Solution

From Butterworth nomographs for $A_{max} = 3$ dB, $A_{min} = 35$ dB, and $f_s/f_1 = 4$, we find $n = 3$, hence $n = 4$ (only even number).

$$K_1 = K_2 = \sqrt{K} = \sqrt{5} = 2.236$$

From the Butterworth coefficients of Appendix C, we find:
First stage

$$a = 1.848 \qquad b = 1.000$$

$$R_{1n} = R_{3n} = 1 \ \Omega$$

$$R_{2n} = KR_{1n} = 2.236 \times 1 \ \Omega = 2.236 \ \Omega$$

$$C_{1n} = \frac{2K+1}{aK} = \frac{2 \times 2.236+1}{1.848 \times 2.236} = \frac{5.472}{4.132} = 1.324 \text{ F}$$

$$C_{2n} = \frac{a}{(2K+1)b} = \frac{1.848}{(2 \times 2.236+1) \times 1} = \frac{1.848}{5.472} = 0.338 \text{ F}$$

Second stage

$$a = 0.765 \qquad b = 1.000$$

$$R_{1n} = R_{3n} = 1 \ \Omega$$

$$R_{2n} = KR_{1n} = 2.236 \times 1 \ \Omega = 2.236 \ \Omega$$

$$C_{1n} = \frac{2K+1}{aK} = \frac{2 \times 2.236 + 1}{0.765 \times 2.236} = \frac{5.472}{1.711} = 3.199 \ F$$

$$C_{2n} = \frac{a}{(2K+1)b} = \frac{0.765}{(2 \times 2.236 + 1)} = \frac{0.765}{5.472} = 0.140 \ F$$

Denormalization

$$ISF = 10^4$$

$$FSF = \frac{\omega_1}{\omega_n} = 2\pi f_1 = 2\pi \times 10^3$$

$$R_1 = R_3 = ISF \times R_n = 10^4 \times 1 \ \Omega = 10 \ k\Omega$$

$$R_2 = ISF \times R_{2n} = 10^4 \times 2.236 \ \Omega \cong 22.4 \ k\Omega$$

First stage

$$C_1 = \frac{C_{1n}}{ISF \times FSF} = \frac{1.324}{2\pi \times 10^7} = 21.1 \ nF$$

$$C_2 = \frac{C_{2n}}{ISF \times FSF} = \frac{0.388}{2\pi \times 10^7} = 5.4 \ nF$$

Second stage

$$C_1 = \frac{C_{1n}}{ISF \times FSF} = \frac{3.199}{2\pi \times 10^7} = 50.9 \ nF$$

$$C_2 = \frac{C_{2n}}{ISF \times FSF} = \frac{0.140}{2\pi \times 10^7} = 2.2 \ nF$$

Figure 3.8a shows the designed filter, and its frequency response is shown in Figure 3.8b.

EXAMPLE 3.6

Design an HP Butterworth filter with the following specifications: 3 dB at 100 Hz and attenuation 30 dB at 50 Hz, with a gain of 5.

Solution

From the Butterworth nomographs for $A_{max} = 3$ dB, $A_{min} = 30$ dB, and for $f_2/f_{s2} = 100/50 = 2$ we find $n = 6$.

$$K_1 = K_2 = K_3 = K^{1/3} = 5^{1/3} = 1.71$$

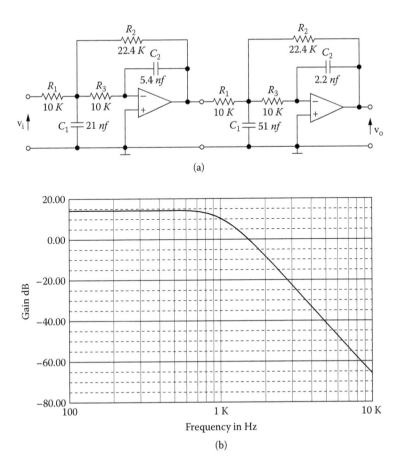

(a)

(b)

FIGURE 3.8 (a) Fourth-order LP Butterworth filter, $f_1 = 1$ kHz, $K = 5$; (b) frequency response.

(a) *Low-pass filter*
First stage

$$a = 1.932 \qquad b = 1.000 \qquad \therefore$$

$$C_{1n} = \frac{2K+1}{aK} = \frac{2 \times 1.71 + 1}{1.932 \times 1.71} = \frac{4.420}{3.304} = 1.338 \text{ F}$$

$$C_{2n} = \frac{a}{(2K+1)b} = \frac{1.932}{4.420} = 0.437 \text{ F}$$

Second stage

$$a = 1.414 \qquad b = 1.000$$

$$C_{1n} = \frac{2K+1}{aK} = \frac{4.420}{1.414 \times 1.71} = \frac{4.420}{2.418} = 1.828 \ \text{F}$$

$$C_{2n} = \frac{a}{(2K+1)b} = \frac{1.414}{4.420} = 0.320 \ \text{F}$$

Third stage

$$a = 0.518 \qquad b = 1.000$$

$$C_{1n} = \frac{2K+1}{aK} = \frac{4.420}{0.518 \times 1.71} = 4.99 \ \text{F}$$

$$C_{2n} = \frac{a}{(2K+1)b} = \frac{0.518}{4.420} = 0.117 \ \text{F}$$

(b) *High-pass filter*
First stage

$$R_{1n} = \frac{1}{C_{1n}} = \frac{1}{1.338} = 0.747 \ \Omega$$

$$R_{2n} = \frac{1}{C_{2n}} = \frac{1}{0.437} = 2.288 \ \Omega$$

Second stage

$$R_{1n} = \frac{1}{C_{1n}} = \frac{1}{1.828} = 0.547 \ \Omega$$

$$R_{2n} = \frac{1}{C_{2n}} = \frac{1}{0.320} = 3.125 \ \Omega$$

Third stage

$$R_{1n} = \frac{1}{C_{1n}} = \frac{1}{4.99} = 0.200 \ \Omega$$

$$R_{2n} = \frac{1}{C_{2n}} = \frac{1}{0.177} = 8.547 \ \Omega$$

$$C_{1n} = C_{3n} = C_n = 1 \ \text{F}$$

$$C_{2n} = \frac{C_n}{K} = \frac{1}{1.71} = 0.585 \ \text{F}$$

Denormalization

$$ISF = 10^4, \; FSF = \frac{\omega_2}{\omega_n} = 2\pi f_2 = 2\pi \times 10^2$$

$$C_1 = C_3 = \frac{C_n}{ISF \times FSF} = \frac{1}{2\pi \times 10^6} \cong 159 \; nF$$

$$C_2 = \frac{C_{2n}}{ISF \times FSF} = \frac{0.585}{2\pi \times 10^6} = 93 \; nF$$

First stage

$$R_1 = ISF \times R_{1n} = 10^4 \times 0.747 \; \Omega \cong 7.5 \; k\Omega$$

$$R_2 = ISF \times R_{2n} = 10^4 \times 2.288 \; \Omega \cong 22.9 \; k\Omega$$

Second stage

$$R_1 = ISF \times R_{1n} = 10^4 \times 0.547 \; \Omega \cong 5.5 \; k\Omega$$

$$R_2 = ISF \times R_{2n} = 10^4 \times 3.125 \; \Omega \cong 31.3 \; k\Omega$$

Third stage

$$R_1 = ISF \times R_{1n} = 10^4 \times 0.200 \; \Omega = 2 \; k\Omega$$

$$R_2 = ISF \times R_{2n} = 10^4 \times 8.547 \; \Omega \cong 85.5 \; k\Omega$$

Figure 3.9a shows the designer filter, and its frequency response is shown in Figure 3.9b.

EXAMPLE 3.7

Design an HP Chebyshev with the following specifications: 2 dB at 500 Hz and attenuation 40 dB at 200 Hz, with a gain of 1.

Solution

From the Chebyshev nomographs for $A_{max} = 2$ dB, $A_{min} = 40$ dB, and $f_2/f_{s2} = 500/200 = 2.5$, we find $n = 4$.
(a) *Low-pass filter*
First stage
From Chebyshev 2-dB coefficients, we find:

$$a = 0.506 \qquad b = 0.222 \qquad \therefore$$

$$C_{1n} = \frac{2K+1}{aK} = \frac{3}{0.506} = 5.926 \; F$$

$$C_{2n} = \frac{a}{(2K+1)b} = \frac{0.506}{3 \times 0.222} = 0.760 \; F$$

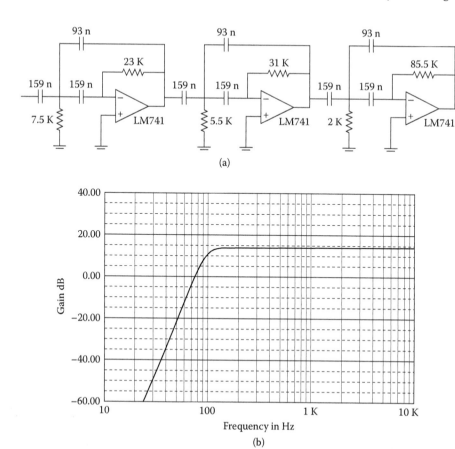

(a)

(b)

FIGURE 3.9 (a) Sixth-order HP Butterworth filter, $f_2 = 100$ Hz, $K = 5$; (b) frequency response.

Second stage

$$a = 0.210 \qquad b = 0.929 \qquad \therefore$$

$$C_{1n} = \frac{2K+1}{aK} = \frac{3}{0.210} = 14.286 \text{ F}$$

$$C_{2n} = \frac{a}{(2K+1)b} = \frac{0.210}{3 \times 0.929} = 0.075 \text{ F}$$

(b) *High-pass filter*
First stage

$$R_{1n} = \frac{1}{C_{1n}} = \frac{1}{5.926} = 0.169 \ \Omega$$

$$R_{2n} = \frac{1}{C_{2n}} = \frac{1}{0.760} = 1.316 \ \Omega$$

Second stage

$$R_{1n} = \frac{1}{C_{1n}} = \frac{1}{14.286} = 0.070 \ \Omega$$

$$R_{2n} = \frac{1}{C_{2n}} = \frac{1}{0.075} = 13.271 \ \Omega$$

$$C_{1n} = C_{3n} = C_n = 1 \ \text{F}$$

$$C_{2n} = \frac{C_n}{K} = \frac{1}{1} = 1 \ \text{F}$$

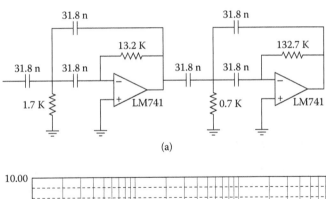

(a)

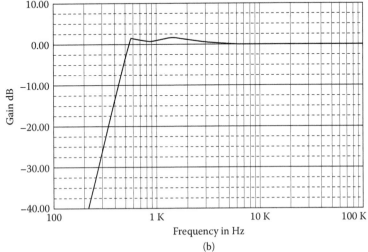

Frequency in Hz

(b)

FIGURE 3.10 (a) Fourth-order HP Chebyshev 2 dB, $f_2 = 500$ Hz, $K = 1$; (b) frequency response.

Denormalization

$$ISF = 10^4, \ FSF = \frac{\omega_2}{\omega_n} = 2\pi f_2 = 2\pi \times 500 = \pi \times 10^3$$

First stage

$$R_1 = ISF \times R_{1n} = 10^4 \times 0.169 \ \Omega \cong 1.7 \ k\Omega$$

$$R_2 = ISF \times R_{2n} = 10^4 \times 1.316 \ \Omega \cong 13.2 \ k\Omega$$

Second stage

$$R_1 = ISF \times R_{1n} = 10^4 \times 0.070 \ \Omega = 0.7 \ k\Omega$$

$$R_2 = ISF \times R_{2n} = 10^4 \times 13.271 \ \Omega = 132.7 \ k\Omega$$

$$C_1 = C_2 = C_3 = \frac{C_n}{ISF \times FSF} = \frac{1}{\pi \times 10^7} = 31.8 \ nF$$

Figure 3.10a shows the designed filter, and its frequency response is shown in Figure 3.10b.

3.4 BAND-PASS FILTERS

When the separation between the upper and lower cutoff frequencies exceeds a ratio of approximately 2, the band-pass filter is considered a **wide-band** type of filter. The specification is then separated into individual low-pass and high-pass requirements and met by a cascade of active low-pass and high-pass filters.

EXAMPLE 3.8

Design a Butterworth band-pass filter with the following specifications: 3 dB at 200 to 800 Hz and attenuation 20 dB below 50 Hz and above 3200 Hz, with gain 1.

Solution

Because the ratio of upper cutoff frequency to lower cutoff frequency is well in excess of an octave, the design will be treated as a cascade of low-pass and high-pass filters. The frequency response requirements can be restated as the following set of individual low-pass and high-pass specifications:

High-pass filter	Low-pass filter
3 dB at 200 Hz	3 dB at 800 Hz
20 dB at 50 Hz	20 dB at 3200 Hz

(a) *Low-pass filter*
From the Butterworth nomographs for $A_{max} = 3$ dB, $A_{min} = 50$ dB, and $f_1/f_{s1} = 3200/800 = 4$, we find $n = 2$. From Butterworth coefficients, we have:

$$a = 1.414 \qquad b = 1.000$$

Hence,

$$R_{1n} = 1\ \Omega \qquad R_{2n} = \frac{R_{1n}}{K} = \frac{1}{1} = 1\ \Omega$$

$$C_{1n} = \frac{2K+1}{aK} = \frac{3}{1.414} = 2.122\ F$$

$$C_{2n} = \frac{a}{(2K+1)b} = \frac{1.414}{3} = 0.471\ F$$

Denormalization

$$ISF = 10^4,\ FSF = \frac{\omega_1}{\omega_n} = 2\pi f_1 = 2\pi \times 800 = 1.6\pi \times 10^3$$

$$R_1 = R_2 = ISF \times R_{1n} = 10^4 \times 1\ \Omega = 10\ k\Omega$$

$$C_1 = \frac{C_{1n}}{ISF \times FSF} = \frac{2.122}{1.6\pi \times 10^7} = 42.2\ nF$$

$$C_2 = \frac{C_{2n}}{ISF \times FSF} = \frac{0.471}{1.6\pi \times 10^7} = 9.4\ nF$$

(b) *High-pass filter*

$$R_{1n} = \frac{1}{C_{1n}} = \frac{1}{2.122} = 0.471\ \Omega$$

$$R_{2n} = \frac{1}{C_{2n}} = \frac{1}{0.471} = 2.122\ \Omega$$

$$C_{1n} = C_{2n} = C_{3n} = 1\ F$$

Denormalization

$$ISF = 10^4,\ FSF = \frac{\omega_2}{\omega_n} = 2\pi f_2 = 2\pi \times 200 = 400\pi$$

$$C_1 = C_2 = C_3 = C = \frac{C_n}{ISF \times FSF} = \frac{1}{4\pi \times 10^6} = 79.6\ nF$$

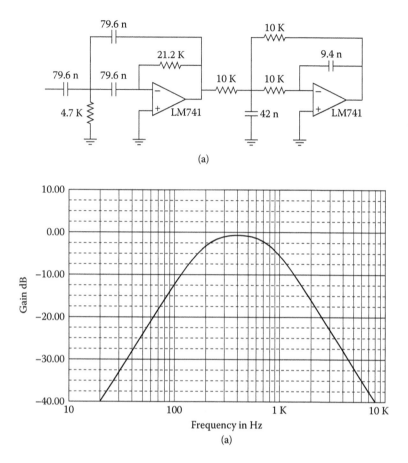

(a)

FIGURE 3.11 (a) Wide-band band-pass Butterworth filter 200 to 800 Hz, $K = 1$; (b) frequency response.

$$R_1 = ISF \times R_{1n} = 10^4 \times 0.471 \; \Omega \cong 4.7 \; k\Omega$$

$$R_2 = ISF \times R_{2n} = 10^4 \times 2.122 \; \Omega \cong 21.2 \; k\Omega$$

Figure 3.11a shows the designed filter, and its frequency response is shown in Figure 3.11b.

3.4.1 NARROW-BAND BAND-PASS FILTER

The basic circuit of the second-order infinite gain MFB narrow-band band-pass filter is shown in Figure 3.12. From this figure, we have:
node v_1

$$-G_1V_i + (G_1 + G_3 + sC_1 + sC_2)V_1 - sC_1V_2 - sC_1V_0 = 0 \qquad (3.18)$$

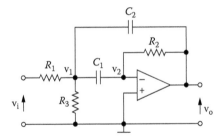

FIGURE 3.12 Second-order multifeedback narrow-band band-pass filter.

node v_2

$$-sC_1V_1 + (sC_1 + G_2)V_2 - G_2V_0 = 0 \tag{3.19}$$

$$V_2 \cong 0 \tag{3.20}$$

From Equations (3.20) and (3.19), we have:

$$V_1 = -\frac{G_2}{sC_1}V_0 \tag{3.21}$$

From Equations (3.18), (3.20), and (3.21), we have:

$$\left\{\frac{G_2[G_1 + G_3 + s(C_1 + C_2)]}{sC_1} + sC_2\right\}V_0 = -G_1V_i \quad \therefore$$

$$\left\{[G_1 + G_3 + s(C_1 + C_2)]G_2 + s^2C_1C_2\right\}V_0 = -sC_1G_1V_i \quad \therefore$$

$$[s^2C_1C_2 + s(C_1 + C_2)G_2 + (G_1 + G_3)G_2]V_0 = -sC_1G_1V_i \quad \therefore$$

$$H(s) = \frac{V_0}{V_i} = -\frac{sC_1G_1}{s^2C_1C_2 + s(C_1 + C_2)G_2 + (G_1 + G_3)G_2} \tag{3.22}$$

For $C_1 = C_2 = C$, we have:

$$H(s) = -\frac{sCG_1}{s^2C^2 + 2sCG_2 + (G_1 + G_3)G_2}$$

$$H(s) = -\frac{\dfrac{G_1}{C}s}{s^2 + \dfrac{2G_2}{C}s + \dfrac{(G_1 + G_3)G_2}{C^2}} \tag{3.23}$$

Hence,

$$H(s) = -\frac{\dfrac{G_1}{C}s}{s^2 + as + \omega_0^2}$$

(3.24)

where

$$\omega_0^2 = \frac{(G_1 + G_3)G_2}{C^2}$$

(3.25)

and

$$a = \frac{1}{Q} = \frac{2G_2}{C} = \frac{2}{R_2 C}$$

(3.26)

For $s = j\omega_0$ $\therefore$

$$K = H(j\omega_0) = -\frac{\dfrac{G_1}{C}j\omega_0}{-\omega_0^2 + j\dfrac{1}{Q}\omega_0 + \omega_0^2} = -\frac{QG_1}{C}   \therefore$$

$$K = -\frac{C}{2G_2}\cdot\frac{G_1}{C} = -\frac{R_2}{2R_1}$$

(3.27)

3.4.1.1 Design Procedure

For $C = 1$ F   and from Equation (3.26), we have:

$$R_2 = 2Q$$

(3.28)

and from Equation (3.27), we have:

$$R_1 = \frac{Q}{K}$$

(3.29)

From Equation (3.25), for $\omega_0 = 1$ rad/s, we have:

$$(G_1 + G_3)G_2 = 1   \therefore$$

$$\frac{1}{R_1} + \frac{1}{R_3} = R_2   \therefore$$

$$\frac{1}{R_3} = R_2 - \frac{1}{R_1} \tag{3.30}$$

and from Equations (3.30), (3.28), and (3.29), we have:

$$\frac{1}{R_3} = 2Q - \frac{K}{Q} \quad \therefore$$

$$\frac{1}{R_3} = \frac{2Q^2 - K}{Q} \quad \therefore$$

$$R_3 = \frac{Q}{2Q^2 - K} \tag{3.31}$$

From the preceding relationship, we have:

$$2Q^2 > K \tag{3.32}$$

For K as a free parameter

$$K = 2Q^2 \tag{3.33}$$

hence

$$R_3 = \infty \quad (i.e., open circuit) \tag{3.34}$$

3.4.1.2 Frequency Response

From the transfer function, Equation (3.23), we have:

$$H(s) = -\frac{\dfrac{G_1}{C}s}{s^2 + \dfrac{2G_2}{C}s + \omega_0^2} = -\frac{\dfrac{G_1}{\omega_0 C}\left(\dfrac{s}{\omega_0}\right)}{\left(\dfrac{s}{\omega_0}\right)^2 + \dfrac{2G_2}{\omega_0 C}\left(\dfrac{s}{\omega_0}\right) + 1} \quad \therefore$$

$$H(s) = -\frac{K_0\left(\dfrac{s}{\omega_0}\right)}{\left(\dfrac{s}{\omega_0}\right)^2 + \dfrac{1}{Q}\left(\dfrac{s}{\omega_0}\right) + 1} \tag{3.35}$$

(1) For $\dfrac{s}{\omega_0} \ll 1$ $\quad \therefore$

$$H(s) \cong -K_0\left(\dfrac{s}{\omega_0}\right) \quad \therefore \quad H(j\omega_0) = -K_0\left(j\dfrac{\omega}{\omega_0}\right) \quad \therefore$$

$$A = 20\log|H(j\omega_0)| = 20\log K_0\left(\dfrac{\omega}{\omega_0}\right) \quad \therefore$$

$$A = 20\log K_0 + 20\log\left(\dfrac{\omega}{\omega_0}\right)$$

Hence, we have:
   slope $= 20$ dB/dec

(2) For $\dfrac{s}{\omega_0} \gg 1$ $\quad \therefore$

$$H(s) \cong -\dfrac{K_0\left(\dfrac{s}{\omega_0}\right)}{\left(\dfrac{s}{\omega_0}\right)^2} = -K_0\left(\dfrac{s}{\omega_0}\right)^{-1} \quad \therefore$$

$$H(j\omega) = -K_0\left(j\dfrac{\omega}{\omega_0}\right)^{-1} \quad \therefore$$

$$A = 20\log|H(j\omega)| = 20\log K_0\left(\dfrac{\omega}{\omega_0}\right)^{-1}$$

$$A = 20\log K_0 - 20\log\left(\dfrac{\omega}{\omega_0}\right)$$

Hence, the slope is -20 dB/dec.

(3) For $\dfrac{s}{\omega_0} = 1$ $\quad \therefore$

$$H(j\omega) = -\dfrac{jK_0}{-1 + j\dfrac{1}{Q} + 1} = -QK_0 = K \quad \therefore$$

$$K_0 = \dfrac{K}{Q} \tag{3.36}$$

Figure 3.13 shows the frequency response of the filter.

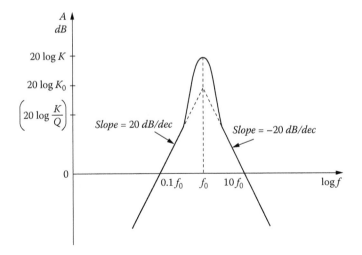

FIGURE 3.13 Frequency response of band-pass filter.

EXAMPLE 3.9

Design a narrow-band band-pass filter with $f_0 = 1$ kHz, $Q = 7$, and $K = 10$.

Solution

$$R_{1n} = \frac{Q}{K} = \frac{7}{10} = 0.7 \ \Omega$$

$$R_{2n} = 2Q = 2 \times 7 = 14 \ \Omega$$

$$R_{3n} = \frac{Q}{2Q^2 - K} = \frac{7}{2 \times 7^2 - 10} = \frac{7}{98 - 10} = 0.08 \ \Omega$$

Denormalization

$$ISF = 10^4, \ FSF = \frac{\omega}{\omega_0} = 2\pi f_0 = 2\pi \times 10^3$$

$$C = \frac{C_n}{ISF \times FSF} = \frac{1}{2\pi \times 10^7} = 15.92 \ \text{nF}$$

$$R_1 = ISF \times R_{1n} = 10^4 \times 0.7 \ \Omega = 7 \ \text{k}\Omega$$

$$R_2 = ISF \times R_{2n} = 10^4 \times 14 \ \Omega = 140 \ \text{k}\Omega$$

$$R_3 = ISF \times R_{3n} = 10^4 \times 0.08 \ \Omega = 800 \ \Omega$$

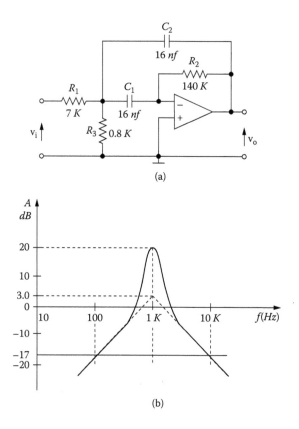

(a)

(b)

FIGURE 3.14 (a) Band-pass filter, $f_0 = 1$ kHz, $Q = 7$, $K = 10$; (b) frequency response.

Figure 3.14a shows the designed filter, and its frequency response is shown in Figure 3.14b.

A much sharper band-pass filter may be obtained by cascading two or more identical band-pass second-order filters. If Q_1 is the quality factor of a single stage and there are n stages, the Q of the filter is expressed by the following equation:

$$Q = \frac{Q_1}{\sqrt{\sqrt[n]{2} - 1}} \tag{3.37}$$

The value of Q_1 and corresponding bandwidths are shown for $n = 1, 2, 3, 4, 5,$ 6, and 7 in Table 3.1, where BW_1 is the bandwidth of a filter with a single stage.

EXAMPLE 3.10

Design a narrow-band band-pass filter with $f_0 = 1$ kHz, $Q = 10$, and with gain as a free parameter.

TABLE 3.1
Bandwidth (BW) and Q for identical cascaded second-order band-pass filter

n	Q	BW
1	Q_1	BW_1
2	$1.554Q_1$	$0.664BW_1$
3	$1.961Q_1$	$0.510BW_1$
4	$2.299Q_1$	$0.435BW_1$
5	$2.593Q_1$	$0.386BW_1$
6	$2.858Q_1$	$0.350BW_1$
7	$3.100Q_1$	$0.323BW_1$

Solution

$$K = 2Q^2 = 2 \times 10^2 = 200 \quad \text{or} \quad 46 \text{ dB}$$

$$R_{1n} = \frac{Q}{K} = \frac{10}{200} = 0.05 \ \Omega$$

$$R_{2n} = 2Q = 2 \times 10 = 20 \ \Omega$$

Denormalization

$$ISF = 10^4, \ FSF = \frac{\omega_0}{\omega_n} = 2\pi f_0 = 2\pi \times 10^3$$

Hence,

$$C = \frac{C_n}{ISF \times FSF} = \frac{1}{2\pi \times 10^7} = 15.9 \ \text{nF}$$

$$R_1 = ISF \times R_{1n} = 10^4 \times 0.05 \ \Omega = 0.5 \ \text{k}\Omega$$

$$R_2 = ISF \times R_{2n} = 10^4 \times 20 \ \Omega = 200 \ \text{k}\Omega$$

Figure 3.15a shows the designed filter, and its frequency response is shown in Figure 3.15b.

EXAMPLE 3.11

Design a sixth-order (three-pole) narrow-band band-pass filter with $f_0 = 750$ Hz, $Q = 8.53$, and $K = 6$.

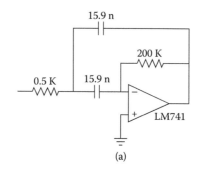

(a)

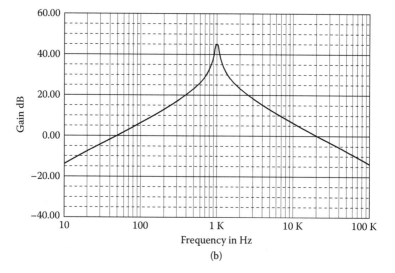

Frequency in Hz

(b)

FIGURE 3.15 (a) Band-pass filter, $f_0 = 1$ kHz, $Q = 10$, K: free parameter; (b) frequency response.

Solution

We will use three identical stages.

$$Q_1 = 0.510, \quad Q = 0.510 \times 8.53 = 4.35$$

$$K_1 = K^{1/3} = 6^{1/3} = 1.82$$

$$R_{1n} = \frac{Q_1}{K_1} = \frac{4.35}{1.82} = 2.39 \ \Omega$$

$$R_{2n} = 2Q = 2 \times 4.35 \ \Omega = 8.7 \ \Omega$$

$$R_{3n} = \frac{Q}{2Q^2 - K_1} = \frac{4.35}{2 \times 4.35^2 - 1.82} = \frac{4.35}{37.845 - 1.82} = \frac{4.35}{36.025} = 0.121 \ \Omega$$

Denormalization

$$ISF = 10^4, \ FSF = \frac{\omega_0}{\omega_n} = 2\pi f_0 = 2\pi \times 750 = 1.5\pi \times 10^3$$

$$C = \frac{C_n}{ISF \times FSF} = \frac{1}{1.5\pi \times 10^7} = 21.2 \ \text{nF}$$

$$R_1 = ISF \times R_{1n} = 10^4 \times 2.39 \ \Omega = 23.9 \ \text{k}\Omega$$

$$R_2 = ISF \times R_{2n} = 10^4 \times 8.7 \ \Omega = 87 \ \text{k}\Omega$$

$$R_3 = ISF \times R_{3n} = 10^4 \times 0.121 \ \Omega = 1.21 \ \text{k}\Omega$$

Figure 3.16a shows the designed filter, and its frequency response is shown in Figure 3.16b.

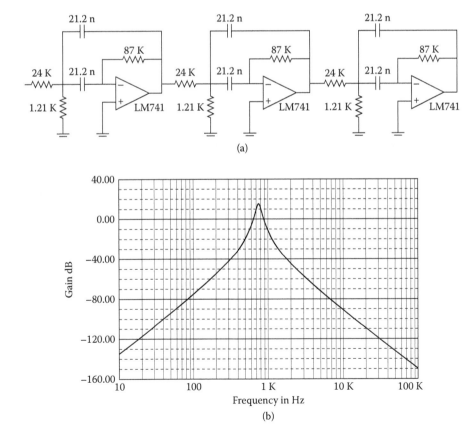

(a)

(b)

FIGURE 3.16 (a) Sixth-order band-pass filter, $f_0 = 750$ Hz, $Q = 8.53$, $K = 6$; (b) frequency response.

3.4.2 Narrow-Band Band-Pass Filter with Two Op-Amps

A high-Q band-pass filter cannot be designed with a single op-amp. A circuit using two op-amps, which can obtain Q values of around 50, is the circuit shown in Figure 3.17.

From this figure, we have:

node v_1

$$-GV_i + (G + G_1 + G_2 + 2sC)V_1 - sCV_2 - sCV_{01} - G_2V_0 = 0 \qquad (3.38)$$

node v_2

$$-sCV_1 + (G + sC)V_2 - GV_{01} = 0 \qquad (3.39)$$

$$V_{01} = \frac{V_0}{K} \qquad (3.40)$$

$$K = \frac{R_4}{R_1} \qquad (3.41)$$

$$V_2 \cong 0 \qquad (3.42)$$

From Equations (3.39), (3.40), and (3.42), we have:

$$V_1 = \frac{G}{sCK}V_0 \qquad (3.43)$$

From Equations (3.38), (3.40), (3.42), and (3.43), we have:

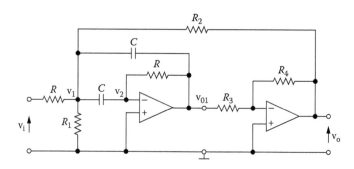

FIGURE 3.17 A second-order band-pass filter with two op-amps.

$$(G + G_1 + G_2 + 2sC)\frac{G}{sCK}V_0 + \frac{sC}{K}V_0 - G_2V_0 = GV_i \qquad \therefore$$

$$[G(G + G_1 + G_2 + 2sC) + s^2C^2 - sCG_2K]V_0 = sCGKV_i \qquad \therefore$$

$$H(s) = \frac{V_0}{V_i} = \frac{sCGK}{s^2C^2 + (2G - KG_2)sC + G(G + G_1 + G_2)} \qquad \therefore$$

$$H(s) = \frac{\dfrac{KG}{C}s}{s^2 + s\dfrac{2G - KG_2}{C} + \dfrac{G(G + G_1 + G_2)}{C^2}} \qquad \therefore$$

$$H(s) = \frac{s\dfrac{KG}{C}}{s^2 + \dfrac{2G - KG_2}{C}s + \omega_0^2} \qquad \therefore$$

where

$$\omega_0^2 = \frac{G(G + G_1 + G_2)}{C^2} \tag{3.44}$$

Hence,

$$H(s) = \frac{\dfrac{KG}{\omega_0 C}\left(\dfrac{s}{\omega_0}\right)}{\left(\dfrac{s}{\omega_0}\right)^2 + \dfrac{2G - KG_2}{\omega_0 C}\left(\dfrac{s}{\omega_0}\right) + 1} \qquad \therefore$$

$$H(s) = \frac{\dfrac{KG}{\omega_0 C}\left(\dfrac{s}{\omega_0}\right)}{\left(\dfrac{s}{\omega_0}\right)^2 + \dfrac{1}{Q}\left(\dfrac{s}{\omega_0}\right) + 1} \tag{3.45}$$

where

$$a = \frac{1}{Q} = \frac{2G - KG_2}{\omega_0 C} \tag{3.46}$$

For $s = j\omega_0$, we have:

$$K_0 = H(j\omega_0) = \frac{j\dfrac{KG}{\omega_0 C}}{-1 + j\dfrac{1}{Q} + 1} = \frac{KQG}{\omega_0 C} = \frac{KG}{\omega_0 C} \cdot \frac{\omega_0 C}{2G - KG_2} \qquad \therefore$$

$$K_0 = \frac{KG}{2G - KG_2} \tag{3.47}$$

From Equation (3.46), we have:

$$KG_2 = 2G - \frac{\omega_0 C}{Q} \qquad \therefore$$

$$\frac{K}{R_2} = \frac{2}{R} - \frac{\omega_0 C}{Q} \qquad \therefore$$

$$R_2 = \frac{KQR}{2Q - \omega_0 RC} \tag{3.48}$$

We put

$$\omega_0 RC = \sqrt{Q} \tag{3.49}$$

From Equations (3.48) and (3.49), we have:

$$R_2 = \frac{KQR}{2Q - \sqrt{Q}} = \frac{KQR}{\sqrt{Q}\left(\frac{2Q}{\sqrt{Q}} - 1\right)} \qquad \therefore$$

$$R_2 = \frac{K\sqrt{Q}R}{2\sqrt{Q} - 1} \tag{3.50}$$

From Equation (3.44), we have:

$$(\omega_0 C)^2 = \frac{1}{R}\left(\frac{1}{R} + \frac{1}{R_1} + \frac{1}{R_2}\right) \qquad \therefore$$

$$(\omega_0 C)^2 R = \frac{1}{R} + \frac{1}{R_1} + \frac{1}{R_2} \qquad \therefore$$

$$\frac{1}{R_1} = \frac{\omega_0^2 R^2 C^2 - 1}{R} - \frac{1}{R_2} \qquad \therefore$$

$$\frac{1}{R_1} = \frac{\omega_0^2 R^2 C^2 - 1}{R} - \frac{2\sqrt{Q} - 1}{KR\sqrt{Q}} \qquad \therefore$$

$$\frac{1}{R_1} = \frac{1}{R}\left(Q - 1 - \frac{2}{K} + \frac{1}{K\sqrt{Q}}\right) \tag{3.51}$$

3.4.2.1 Design Procedure

For $\omega_0 = 1$ rad/s (normalized), we accept

$$C_n = 1 \text{ F}, \quad R_{3n} = 1 \ \Omega \tag{3.52}$$

From Equation (3.49), we have:

$$R = \sqrt{Q} \tag{3.53}$$

$$\frac{1}{R_1} = \frac{1}{R}\left(Q - 1 - \frac{2}{K} + \frac{1}{K\sqrt{Q}}\right) \tag{3.54}$$

$$R_2 = \frac{KQ}{2\sqrt{Q} - 1} \tag{3.55}$$

EXAMPLE 3.12

Design a second-order narrow-band BPF with two op-amps with a center frequency of 1 kHz, $Q = 40$, and $K = 5$.

Solution

$$R_n = \sqrt{Q} = \sqrt{40} = 6.325 \ \Omega$$

$$R_{3n} = 1 \ \Omega$$

$$R_{4n} = K = 5 \ \Omega$$

$$R_{2n} = \frac{5\sqrt{40} \times 6.325}{2 \times 6.325 - 1} = \frac{200}{11.65} = 17.2 \ \Omega$$

$$\frac{1}{R_{1n}} = \frac{1}{6.325}\left(40 - 1 - \frac{2}{5} - \frac{1}{5 \times 6.325}\right) = 0.164 \ \Omega$$

Denormalization

$$ISF = 10^4, \quad FSF = \frac{\omega_0}{\omega_n} = 2\pi f_0 = 2\pi \times 10^3$$

$$R = ISF \times R_n = 10^4 \times 6.325 \ \Omega = 63.3 \ \text{k}\Omega$$

$$R_1 = ISF \times R_{1n} = 10^4 \times 0.164 \ \Omega = 1.64 \ \text{k}\Omega$$

$$R_2 = ISF \times R_{2n} = 10^4 \times 17.2 \ \Omega = 172 \ \text{k}\Omega$$

$$R_3 = ISF \times R_{3n} = 10^4 \times 1 \ \Omega = 10 \ k\Omega$$

$$R_4 = ISF \times R_{4n} = 10^4 \times 5 \ \Omega = 50 \ k\Omega$$

$$K_0 = \frac{K}{2 - K\dfrac{G_2}{G}} = \frac{K}{2 - K\dfrac{R}{R_2}} = \frac{5}{2 - 5\dfrac{63.3}{172}} = 31.3 \quad \text{or} \quad 29.9 \ \text{dB}$$

$$C = \frac{1}{2\pi \times 10^7} = 15.92 \ \text{nF}$$

Figure 3.18 shows the designed filter with its frequency response.

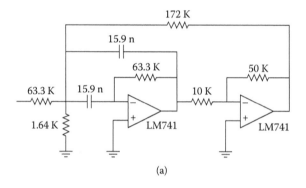

(a)

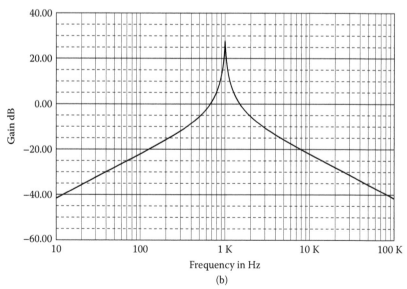

Frequency in Hz

(b)

FIGURE 3.18 (a) Second-order narrow-band BPF, $f_0 = 1$ kHz, $Q = 40$, $K = 5$; (b) frequency response.

3.4.3 DELIYANNIS'S BAND-PASS FILTER

The active filter shown in Figure 3.12 appears to be a good filter for $Q \leq 10$. Deliyannis has proposed the band-pass filter shown in Figure 3.19 in which the Q value can be very large.

From this figure, we have:

node v_1

$$-\frac{1}{R}V_i + \left(\frac{1}{R} + 2sC\right)V_1 - sCV_2 - sCV_0 = 0 \tag{3.56}$$

node v_2

$$-sCV_1 + \left(\frac{1}{kR} + sC\right)V_2 - \frac{1}{kR}V_0 = 0 \tag{3.57}$$

node v_3

$$V_3 = \frac{R_a}{R_a + R_b}V_0 = \beta V_0 \tag{3.58}$$

where

$$\beta = \frac{R_a}{R_a + R_b} = \frac{r}{r + Mr} = \frac{1}{1 + M} \tag{3.59}$$

For an ideal op-amp, we have:

$$V_3 - V_2 = 0 \qquad \therefore$$

$$V_2 = V_3 = \beta V_0 \tag{3.60}$$

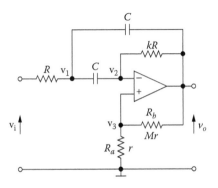

FIGURE 3.19 Deliyannis's band-pass filter.

From Equations (3.60) and (3.57), we have:

$$sCV_1 = \left(\frac{1}{kR} + sC\right)\beta V_0 - \frac{1}{kR}V_0 \quad \therefore$$

$$V_1 = \frac{(1+skRC)\beta - 1}{skRC}V_0 \quad \therefore$$

$$V_1 = -\frac{(1-\beta) - \beta kRCs}{skRC}V_0 \tag{3.61}$$

From Equations (3.59) and (3.56), we have:

$$(1+2sRC)\left[-\frac{(1-\beta)-\beta kRCs}{skRC}\right]V_0 - sRC\beta V_0 - sRCV_0 = V_i \quad \therefore$$

$$\{(1+2sRC)[(1-\beta)-\beta kRCs] + s^2R^2C^2k\beta + s^2R^2C^2k\}V_0 = sRCkV_i \quad \therefore$$

$$[(1-\beta)-\beta kRCs + 2(1-\beta)RCs - s^2R^2C^2k\beta + s^2R^2C^2k]V_0 = -sRCkV_i \quad \therefore$$

$$H(s) = \frac{V_0}{V_i} = \frac{skRC}{s^2R^2C^2k(1-\beta) + RC[2(1-\beta)-k\beta]s + (1-\beta)} \quad \therefore$$

$$H(s) = \frac{-\dfrac{s}{(1-\beta)RC}}{s^2 + \dfrac{2(1-\beta)-\beta k}{(1-\beta)kRC}s + \dfrac{1}{kR^2C^2}} \quad \therefore$$

$$\therefore$$

$$H(s) = \frac{-\dfrac{s}{(1-\beta)RC}}{s^2 + \dfrac{2(1-\beta)-\beta k}{(1-\beta)kRC}s + \omega_0^2} \tag{3.62}$$

where

$$\omega_0^2 = \frac{1}{kR^2C^2} \quad \therefore$$

$$\omega_0 = \frac{1}{RC\sqrt{k}} \tag{3.63}$$

Hence

$$H(s) = \cfrac{-\cfrac{1}{(1-\beta)\omega_0 RC}\left(\cfrac{s}{\omega_0}\right)}{\left(\cfrac{s}{\omega_0}\right)^2 + \cfrac{2(1-\beta)-\beta k}{(1-\beta)kRC\omega_0}\left(\cfrac{s}{\omega_0}\right) + 1} \qquad \therefore$$

$$H(s) = -\cfrac{\cfrac{1}{(1-\beta)\omega_0 RC}\left(\cfrac{s}{\omega_0}\right)}{\left(\cfrac{s}{\omega_0}\right)^2 + \cfrac{1}{Q}\left(\cfrac{s}{\omega_0}\right) + 1} \qquad (3.64)$$

where

$$Q = \frac{(1-\beta)kRC\omega_0}{2(1-\beta)-\beta k} \qquad (3.65)$$

It is important that Equation (3.65) be well understood. In particular, note that when $2(1-\beta) = \beta k$, the denominator goes to zero, and Q* becomes infinite, which means that the filter oscillates. Thus, given a value of k, which sets f_0 and the minimum Q, we can find the value of β that will give the maximum Q. From Equation (3.65) we can find:

$$\beta_{max} = \frac{2}{2+k} \qquad (3.66)$$

From Equation (3.59), we have:

$$\beta = \frac{1}{1+M} \qquad (3.67)$$

and from Equations (3.67) and (3.65), we have:

$$Q = \cfrac{\left(1-\cfrac{1}{1+M}\right)\sqrt{k}}{2\left(1-\cfrac{1}{1+M}\right)-\cfrac{1}{1+M}k} \qquad \therefore$$

$$Q = \frac{M\sqrt{k}}{2M-k} \qquad \therefore$$

---

* This technique is frequently referred to as Q **enhancement**.

$$M = \frac{kQ}{2Q - \sqrt{k}} \qquad (3.68)$$

From Equation (3.65), we have:

$$2(1-\beta)Q - \beta kQ = (1-\beta)\sqrt{k} \qquad \therefore$$

$$\beta = \frac{2Q - \sqrt{k}}{2Q + kQ - \sqrt{k}} \qquad (3.69)$$

For $s = j\omega_0$, Equation (3.64) becomes:

$$K = H(j\omega_0) = \frac{-\dfrac{1}{(1-\beta)\omega_0 RC}j}{-1 + j\dfrac{1}{Q} + 1} = -\frac{\dfrac{1}{(1-\beta)\omega_0 RC}}{\dfrac{1}{Q}} = -\frac{\dfrac{1}{(1-\beta)\omega_0 RC}}{\dfrac{2(1-\beta) - \beta k}{(1-\beta)kRC\omega_0}} \qquad \therefore$$

$$K = -\frac{k}{2(1-\beta) - \beta k} \qquad (3.70)$$

If we desire unity passband gain ($K = 1$), as is often reasonable, we have to attenuate the input signal by a factor of $1/K$. Here, we will be looking for an attenuator formed from resistors R_1 and R_3, as shown in Figure 3.20.

Hence,

$$\frac{1}{K} = \frac{R_3}{R_1 + R_2} \qquad (3.71)$$

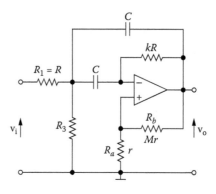

FIGURE 3.20 Deliyannis's band-pass filter with attenuator input R_1, R_3.

$$R = \frac{R_1 R_3}{R_1 + R_3} \tag{3.72}$$

From these equations, we have:

$$R_1 = KR \tag{3.73}$$

$$R_3 = \frac{KR_1}{K-1} \tag{3.74}$$

3.4.3.1 Design Procedure

For $\omega_0 = 1$ rad/s and $C = 1$ F, from Equation (3.63), we have:

$$R = \frac{1}{\sqrt{k}} \tag{3.75}$$

Choose k to be a number that is a perfect square for convenience of calculation (e.g., 4, 9, 16,..., 100, 121,...).

From Equation (3.68) calculate M.

Select resistors $R_a = r$ and $R_b = Mr$ such that their sum is greater than 20 $K\Omega$ and less than 200 $K\Omega$.

If the passband gain K is not going to be a problem, leave out R_3 and make $R_1 = R$. However, if you want to ensure unity passband gain, calculate K from Equations (3.70), (3.67), and (3.68), i.e.,

$$K = \frac{k}{2(1-\beta) - \beta k} = \frac{k}{2\left(1 - \dfrac{1}{1+M}\right) - \dfrac{k}{1+M}} \quad \therefore$$

$$K = \frac{k(1+M)}{2M-k} \quad \therefore$$

$$K = \frac{k\left(1 + \dfrac{kQ}{2Q - \sqrt{k}}\right)}{\dfrac{2kQ}{2Q - \sqrt{k}} - k} = \frac{k\dfrac{2Q - \sqrt{k} + kQ}{2Q - \sqrt{k}}}{\dfrac{2kQ - 2kQ - k\sqrt{k}}{2Q - \sqrt{k}}} \quad \therefore$$

$$K = \frac{2Q - \sqrt{k} + kQ}{\sqrt{k}} = \frac{(2Q - \sqrt{k})\left(1 + \dfrac{kQ}{2Q - \sqrt{k}}\right)}{\sqrt{k}} \quad \therefore$$

$$K = \frac{(1+M)(2Q - \sqrt{k})}{\sqrt{k}} = \frac{1+M}{\sqrt{k}} \cdot \frac{(2Q - \sqrt{k})kQ}{kQ} \qquad \therefore$$

$$K = \frac{1+M}{\sqrt{k}} \cdot \frac{kQ}{M} \qquad \therefore$$

$$K = \frac{Q(1+M)\sqrt{k}}{M} \tag{3.76}$$

EXAMPLE 3.13

Design a filter with $f_0 = 200$ Hz and $Q = 12$.

Solution

We accept $k = 25$ ($\sqrt{k} = 5$).

$$R_n = \frac{1}{\sqrt{k}} = \frac{1}{5} = 0.200 \ \Omega$$

$$M = \frac{kQ}{2Q - \sqrt{k}} = \frac{25 \times 12}{24 - 5} = \frac{300}{19} = 15.8$$

$$K = \frac{Q(1+M)\sqrt{k}}{M} = \frac{12 \times 16.8 \times 5}{15.8} = 63.8 \quad \text{or} \quad 36 \ \text{dB}$$

$$R_{1n} = KR_n = 63.8 \times 0.200 = 12.8 \ \Omega$$

$$R_{2n} = kR_n = 25 \times 0.200 = 5 \ \Omega$$

$$R_{3n} = \frac{KR_n}{K-1} = \frac{63.8 \times 0.2}{63.8 - 1} = 0.203 \ \Omega$$

$$r_n = 1 \ \Omega \qquad \therefore \qquad Mr_n = 15.8 \ \Omega$$

Denormalization

$$ISF = 10^4, \ FSF = \frac{\omega_0}{\omega_n} = \frac{2\pi f_0}{1} = 2\pi \times 200 = 4\pi \times 10^2 \qquad \therefore$$

$$C = \frac{C_n}{ISF \times FSF} = \frac{1}{4\pi \times 10^6} = 79.6 \ \text{nF}$$

$$R = ISF \times R_n = 10^4 \times 0.2 \ \Omega = 2 \ \text{k}\Omega$$

$$R_1 = ISF \times R_{1n} = 10^4 \times 12.8 \ \Omega = 128 \ \text{k}\Omega$$

$$R_2 = ISF \times R_{2n} = 10^4 \times 5 \ \Omega = 50 \ \text{k}\Omega$$

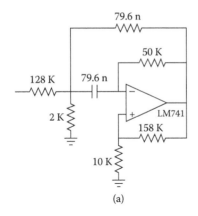

(a)

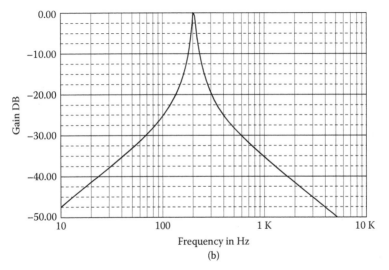

(b)

FIGURE 3.21 (a) Deliyannis's band-pass filter, $f_0 = 200$ Hz, $Q = 12$, $K = 1$; (b) frequency response.

$$R_3 = ISF \times R_{3n} = 10^4 \times 0.203 \ \Omega \cong 2 \ \text{k}\Omega$$

$$r = ISF \times r_n = 10^4 \times 1 \ \Omega = 10 \ \text{k}\Omega$$

$$Mr = ISF \times Mr_n = 10^4 \times 15.8 \ \Omega = 158 \ \text{k}\Omega$$

Figure 3.21a shows the designed filter for $K = 1$, and its frequency response is shown in Figure 3.21b.

Figure 3.22a shows the designed filter with K as a free parameter, and its frequency response is shown in Figure 3.22b.

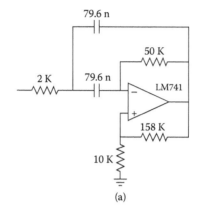

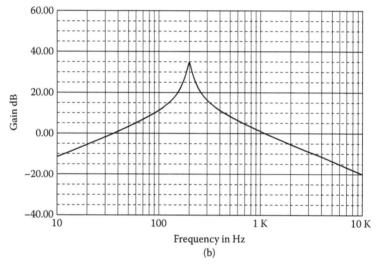

FIGURE 3.22 (a) Deliyannis's band-pass filter, $f_0 = 200$ Hz, $Q = 12$, K: free parameter; (b) frequency response.

3.5 BAND-REJECT FILTERS

3.5.1 WIDE-BAND BAND-REJECT FILTERS

Wide-band band-reject filters can be designed by first separating the specification into individual low-pass and high-pass requirements. Low-pass and high-pass filters are then independently designed and combined by paralleling the inputs and summing both outputs to form the band-reject filters.

A wide-band approach is valid when the separation between cutoffs is an octave or more for all-pole filters so that minimum interaction occurs in the stopband when the outputs are summed (Figure 3.23).

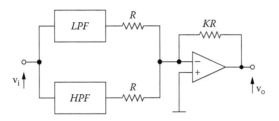

FIGURE 3.23 Wide-band band-reject filters.

EXAMPLE 3.14

Design a wide-band band-reject Butterworth filter having 3 dB at 300 and 900 Hz and greater than 40 dB of attenuation between 750 and 360 Hz, respectively, with a gain of 1.

Solution

As the ratio of upper cutoff to lower cutoff is well in excess of an octave, a wide-band approach can be used. First, separate the specification into individual low-pass and high-pass requirements.

Low-pass	High-pass
3 dB at 300 Hz	3 dB at 900 Hz
40 dB at 750 Hz	40 dB at 360 Hz

(a) *Low-pass filter*
From the Butterworth nomographs for $A_{max} = 3$ dB, $A_{min} = 40$ dB, and $f_s/f_1 = 750/300 = 2.5$, we have $n = 5$; hence, $n = 6$ (only even order).
First stage
From the Butterworth coefficients of Appendix C, we find

$$a = 1.932 \qquad b = 1.000$$

$$R_{1n} = R_{2n} = R_{3n} = 1 \ \Omega \ (K = 1)$$

$$C_{1n} = \frac{2K+1}{aK} = \frac{3}{1.932} = 1.553 \ F$$

$$C_{2n} = \frac{a}{(2K+1)b} = \frac{1.932}{3} = 0.644 \ F$$

Second stage

$$a = 1.414 \qquad b = 1.000$$

$$C_{1n} = \frac{3}{1.414} = 2.122 \ F$$

$$C_{2n} = \frac{1.414}{3} = 0.471 \ F$$

Third stage

$$a = 0.518 \qquad b = 1.000$$

$$C_{1n} = \frac{3}{0.518} = 5.792 \text{ F}$$

$$C_{2n} = \frac{0.518}{3} = 0.173 \text{ F}$$

Denormalization

$$ISF = 10^4, \; FSF = \frac{\omega_1}{\omega_n} = 2\pi f_1 = 2\pi \times 300 = 6\pi \times 10^2$$

$$R_1 = R_2 = R_3 = ISF \times R_{1n} = 10^4 \times 1 \; \Omega = 10 \text{ k}\Omega$$

First stage

$$C_1 = \frac{C_{1n}}{ISF \times FSF} = \frac{1.553}{6\pi \times 10^6} = 82.4 \text{ nF}$$

$$C_2 = \frac{C_{2n}}{ISF \times FSF} = \frac{0.644}{6\pi \times 10^6} = 34.2 \text{ nF}$$

Second stage

$$C_1 = \frac{C_{1n}}{ISF \times FSF} = \frac{2.122}{6\pi \times 10^6} = 112.6 \text{ nF}$$

$$C_2 = \frac{C_{2n}}{ISF \times FSF} = \frac{0.471}{6\pi \times 10^6} = 25 \text{ nF}$$

Third stage

$$C_1 = \frac{C_{1n}}{ISF \times FSF} = \frac{5.792}{6\pi \times 10^6} = 307.3 \text{ nF}$$

$$C_2 = \frac{C_{2n}}{ISF \times FSF} = \frac{0.173}{6\pi \times 10^6} = 9.2 \text{ nF}$$

(b) *High-pass filter*
From the Butterworth nomographs for $A_{max} = 3$ dB, $A_{min} = 40$ dB, and $f_2/f_s = 900/360 = 2.5$, we have $n = 5$; hence, $n = 6$.

$$C_{1n} = C_{2n} = C_{3n} = 1 \text{ F} \; (K = 1)$$

From the preceding low-pass filter, we have:

First stage

$$R_{1n} = \frac{1}{C_{1n}} = \frac{1}{1.553} = 0.644 \ \Omega$$

$$R_{2n} = \frac{1}{C_{2n}} = \frac{1}{0.644} = 1.553 \ \Omega$$

Second stage

$$R_{1n} = \frac{1}{C_{1n}} = \frac{1}{2.122} = 0.471 \ \Omega$$

$$R_{2n} = \frac{1}{C_{2n}} = \frac{1}{0.471} = 2.122 \ \Omega$$

Third stage

$$R_{1n} = \frac{1}{C_{1n}} = \frac{1}{5.792} = 0.173 \ \Omega$$

$$R_{2n} = \frac{1}{C_{2n}} = \frac{1}{0.173} = 5.792 \ \Omega$$

Denormalization

$$ISF = 10^4, \ FSF = \frac{\omega_2}{\omega_n} = 2\pi f_2 = 2\pi \times 900 = 1.8\pi \times 10^3$$

$$C_1 = C_2 = C_3 = \frac{C_n}{ISF \times FSF} = \frac{1}{1.8\pi \times 10^7} = 17.7 \ \text{nF}$$

First stage

$$R_1 = ISF \times R_{1n} = 10^4 \times 0.644 \ \Omega = 6.4 \ \text{k}\Omega$$

$$R_2 = ISF \times R_{2n} = 10^4 \times 1.553 \ \Omega = 15.5 \ \text{k}\Omega$$

Second stage

$$R_1 = ISF \times R_{1n} = 10^4 \times 0.471 \ \Omega = 4.7 \ \text{k}\Omega$$

$$R_2 = ISF \times R_{2n} = 10^4 \times 2.122 \ \Omega = 21.2 \ \text{k}\Omega$$

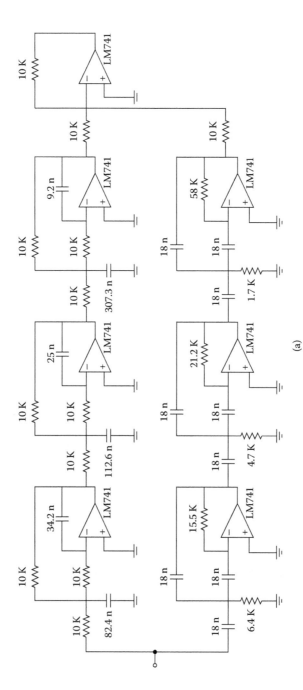

FIGURE 3.24 (a) Wide-band band-reject filter, 300 to 900 Hz, $K = 1$; (b) frequency response.

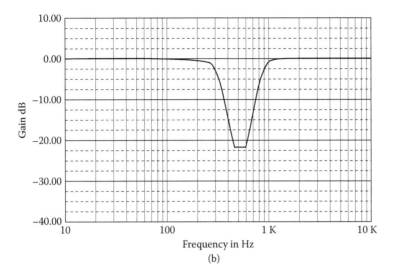

FIGURE 3.24 (Continued)

Third stage

$$R_1 = ISF \times R_{1n} = 10^4 \times 0.173 \ \Omega = 1.7 \ k\Omega$$

$$R_2 = ISF \times R_{2n} = 10^4 \times 5.792 \ \Omega = 57.9 \ k\Omega$$

Figure 3.24a shows the designed filter, and its frequency response is shown in Figure 3.24b.

3.5.2 NARROW-BAND BAND-REJECT FILTER

The circuit of Figure 3.25 is a **narrow-band band-reject** or **notch filter**. Undesired frequencies are attenuated in the stopband. For example, it may be necessary to attenuate 50 Hz, 60 Hz, or 400 Hz noise signals induced in a circuit by motor generators.

From this figure, we have:

node v_1

$$\frac{1}{R_1} V_i + \left(\frac{1}{R_1} + sC_1 + sC_2 \right) V_1 - sC_2 V_2 - sC_1 V_0 = 0 \tag{3.77}$$

node v_2

$$-sC_2 + \left(\frac{1}{R_2} + sC_2 \right) V_2 - \frac{1}{R_2} V_0 = 0 \tag{3.78}$$

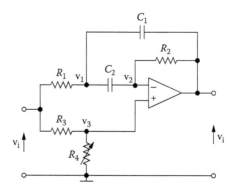

FIGURE 3.25 Narrow-band band-reject or notch filter.

node v_3

$$V_3 = \frac{R_4}{R_3 + R_4}V_i = bV_i \qquad (3.79)$$

where

$$b = \frac{R_4}{R_3 + R_4} \qquad (3.80)$$

From Equation (3.78), we have:

$$V_2 = \frac{V_0 + sR_2C_2V_1}{1 + sR_2C_2} \qquad (3.81)$$

For an ideal op-amp, we have:

$$V_2 = V_3 = bV_i \qquad (3.82)$$

From Equations (3.81), (3.82), and (3.77), we have:

$$(1 + sR_1C_1 + sR_2C_2)V_1 - sR_1C_2V_2 - sR_1C_1V_0 = V_i \qquad \therefore$$

$$(1 + sR_1C_1 + sR_2C_2)\left[\frac{bV_i(1 + sR_2C_2)}{sR_2C_2} - \frac{V_0}{sR_2C_2}\right] - sR_1C_2bV_i - sR_1C_1V_0 = V_i \qquad \therefore$$

$$(1 + sR_1C_1 + sR_1C_2 + s^2R_1R_2C_1C_2)V_0 = (b - sR_2C_2 + sbR_2C_2 + sbR_1C_1$$
$$+ s^2R_1R_2C_1C_2 + sbR_1R_2)V_i \qquad \therefore$$

$$H(s) = \frac{V_0}{V_i} = \frac{s^2 R_1 R_2 C_1 C_2 + s[b(R_1 C_1 + R_2 C_2 + R_1 C_2) - R_2 C_2] + b}{s^2 + R_1 R_2 C_1 C_2 + sR_1 (C_1 + C_2) + 1} \quad (3.83)$$

If we put

$$\omega_0^2 = \frac{1}{R_1 R_2 C_1 C_2} \quad (3.84)$$

and

$$Q = \frac{1}{\omega_0 R_1 (C_1 + C_2)} \quad (3.85)$$

Hence

$$H(s) = \frac{b\left(\frac{s}{\omega_0}\right)^2 + \omega_0[b(R_1 C_1 + R_2 C_2 + R_1 C_2) - R_2 C_2]\left(\frac{s}{\omega_0}\right) + b}{\left(\frac{s}{\omega_0}\right)^2 + \frac{1}{Q}\left(\frac{s}{\omega_0}\right) + 1} \quad (3.86)$$

3.5.2.1 Design Procedure

For $\omega_0 = 1$ rad/s, $C_1 = C_2 = C = 1$ F, and $R_3 = 1\ \Omega$, from Equation (3.85), we have:

$$Q = \frac{1}{2R_1} \quad \therefore$$

$$R_1 = \frac{1}{2Q} \quad (3.87)$$

From Equation (3.84), we have:

$$R_2 = \frac{1}{R_1} = 2Q \quad (3.88)$$

When $s = j\omega_0 = j1$, from Equation (3.86), we have:

$$H(j\omega_0) = \frac{-b + j[b(R_1 + R_2 + R_1) - R_2] + b}{-1 + j\frac{1}{Q} + j} = 0 \quad \therefore$$

$$b = \frac{R_2}{2R_1 + R_2} \quad \text{or}$$

$$\frac{R_4}{R_3 + R_4} = \frac{R_2}{2R_1 + R_2} \quad \therefore$$

$$\frac{R_4}{1 + R_4} = \frac{2Q}{2\dfrac{1}{2Q} + 2Q} = \frac{2Q^2}{1 + 2Q^2} \quad \therefore$$

$$R_4 = 2Q^2 \tag{3.89}$$

EXAMPLE 3.15

Design a notch filter for $f_0 = 50$ Hz and $Q = 5$.

Solution

Let $C_n = 1$ F.

$$R_{3n} = 1 \ \Omega$$

$$R_{1n} = \frac{1}{2Q} = \frac{1}{20} = 0.05 \ \Omega$$

$$R_{2n} = 2Q = 2 \times 10 = 20 \ \Omega$$

$$R_{4n} = 2Q^2 = 2 \times 10^2 = 200 \ \Omega$$

Denormalization
Let $ISF = 10^3$.

$$FSF = \frac{\omega_0}{\omega_n} = 2\pi f_0 = 100\pi$$

$$C = \frac{C_n}{ISF \times FSF} = \frac{1}{\pi \times 10^5} = 3.2 \ \mu F$$

$$R_1 = ISF \times R_{1n} = 10^3 \times 0.05 \ \Omega = 50 \ \Omega$$

$$R_2 = ISF \times R_{2n} = 10^3 \times 20 \ \Omega = 20 \ k\Omega$$

$$R_3 = ISF \times R_{3n} = 10^3 \times 1 \ \Omega = 1 \ k\Omega$$

$$R_4 = ISF \times R_{4n} = 10^3 \times 200 \ \Omega = 200 \ k\Omega$$

Figure 3.26a shows the designed filter, and its frequency response is shown in Figure 3.26b.

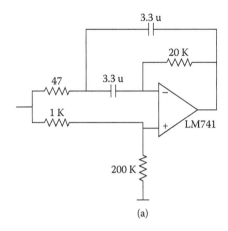

(a)

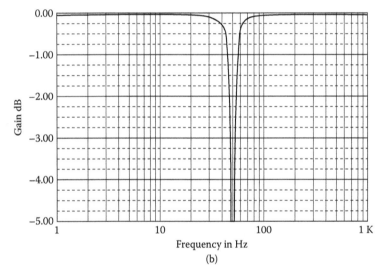

FIGURE 3.26 (a) Notch filter, $f_0 = 50$ Hz, $Q = 5$; (b) frequency response.

3.5.3 MFB NARROW-BAND BAND-REJECT FILTER

By combining the band-pass section with a summing amplifier, the band-reject structure of Figure 3.27 can be derived.

From this figure, we have:

$$V_0 = -V_i \left(\frac{Ks}{s^2 + as + \omega_0^2} + K \right)$$

(3.90)

where

$$K = \frac{R_6}{R_5}$$

(3.91)

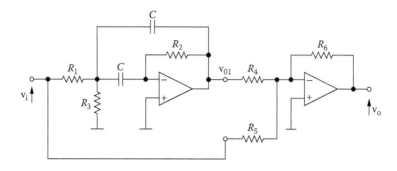

FIGURE 3.27 Multifeedback narrow-band band-reject filter ($Q \le 20$).

and

$$K = \frac{R_2}{2R_1} \qquad (3.92)$$

and for $R_1 = 1\ \Omega$ and $R_4 = R_6 = 1\ \Omega$, hence

$$R_5 = \frac{R_6}{K} \qquad (3.93)$$

and

$$V_0 = -K\left(\frac{s^2 + 1}{s^2 + as + \omega_0^2}\right)V_i \qquad \therefore$$

$$H(s) = -\frac{K(s^2 + 1)}{s^2 + as + \omega_0^2} \qquad (3.94)$$

EXAMPLE 3.16

Design an MFB narrow-band band-reject filter with $f_0 = 1$ kHz, $Q = 6$, and $K = 5$.

Solution

$$R_{1n} = \frac{Q}{K} = \frac{6}{5} = 1.2\ \Omega$$

$$R_{2n} = 2Q = 2 \times 6 = 12\ \Omega$$

$$R_{3n} = \frac{Q}{2Q^2 - K} = \frac{6}{72 - 5} = 0.09\ \Omega$$

$$R_{4n} = R_{6n} = 1\ \Omega$$

$$R_{5n} = \frac{R_{6n}}{K} = \frac{2}{10} = 0.2\ \Omega$$

Denormalization
Let $ISF = 10^4$.

$$FSF = \frac{\omega_0}{\omega_n} = 2\pi f_0 = 2\pi \times 10^3$$

$$C = \frac{C_n}{ISF \times FSF} = \frac{1}{2\pi \times 10^7} = 15.9 \ nF$$

$$R_1 = ISF \times R_{1n} = 10^4 \times 1.2 \ \Omega = 12 \ k\Omega$$

$$R_2 = ISF \times R_{2n} = 10^4 \times 12 \ \Omega = 120 \ k\Omega$$

$$R_3 = ISF \times R_{3n} = 10^4 \times 0.09 \ \Omega = 0.9 \ k\Omega$$

$$R_4 = ISF \times R_{4n} = 10^4 \times 1 \ \Omega = 10 \ k\Omega$$

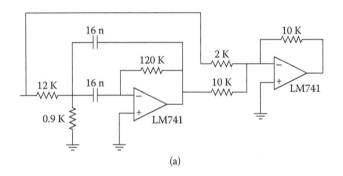

(a)

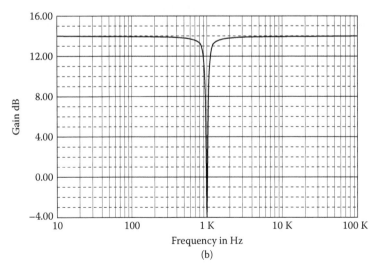

(b)

FIGURE 3.28 (a) Multifeedback narrow-band band-reject filter, $f_0 = 1$ kHz, $Q = 6$, $K = 5$; (b) frequency response.

$$R_5 = ISF \times R_{5n} = 10^4 \times 0.2 \ \Omega = 2 \ k\Omega$$

$$R_6 = ISF \times R_{6n} = 10^4 \times 1 \ \Omega = 10 \ k\Omega$$

Figure 3.28a shows the designed filter, and its frequency response is shown in Figure 3.28b.

3.6 COMMENTS ON MFB FILTERS

3.6.1 LOW-PASS FILTERS

1. In the case of multiple-stage filters ($n > 2$), the K parameter for each stage need not be the same.
2. For best performance, the input resistance of the op-amp should be

$$R_{eq} = R_3 + \frac{R_1 R_2}{R_1 + R_3} \tag{3.95}$$

3. Standard resistance values of 5% tolerance normally yield acceptable results in the lower-order cases. For orders five and six, resistances of 2% tolerance probably should be used, and for order seven, 1% tolerance probably should be used.
4. In the case of capacitors, percentage tolerances should parallel those given earlier for the resistors for best results. As precision capacitors are relatively expensive, it may be desirable to use capacitors of higher tolerances, in which case trimming is generally required. In the case of lower orders ($n \leq 4$), 10% capacitors are quite often satisfactory.
5. The inverting gain of each stage of the filter is R_2/R_1. Gain adjustment can be made by using a potentiometer in lieu of R_2.
6. For minimum dc offset, a resistance equal to R_{eq} of 2 W can be placed in the noninverting input to ground.
7. In a low-pass filter section, maximum gain peaking is very nearly equal to Q at f_1. So, as a rule of thumb:
   (a) Op-amp bandwidth (BW) should be at least

$$BW = 100 \times K \times Q \times f_1 \tag{3.96}$$

   (b) For adequate full-power response, the slew rate (SR) of the op-amp must be

$$SR > \pi \times V_{0pp} \times BW_f \quad V/s \tag{3.97}$$

where BW_f is the filter BW.

EXAMPLE 3.17

A unity-gain 20-kHz 5-pole, 3-db ripple Chebyshev MFB filter has:

$$BW_1 = 100 \times 1 \times 2.139 \times 20 \times 10^3 \cong 4.3 \text{ MHz}$$

$$BW_2 = 100 \times 1 \times 4.795 \times 20 \times 10^3 \cong 17.6 \text{ MHz}$$

$$BW_3 = 50 \times f_1 = 50 \times 20 \times 10^3 = 1 \text{ MHz}$$

where $a = a_i / \sqrt{b_i}$ and $Q = 1/a$ and $BW = 50 \times f_1$ is the bandwidth of the first-order filter.

The op-amp BW = 17.6 MHz (worst case).

$$SR > \pi \times 20 \times 10^3 \ V/s > 1.3 \ V/\mu s$$

3.6.2 HIGH-PASS FILTERS

1. For best performance, the input resistance of the op-amp should be at least 10 times $R_{eq} = R_2$.
2. Standard resistance values of 5% tolerance normally yield acceptable results in the lower-order cases. For orders five and six, resistances of 2% tolerance probably should be used, and for orders seven and eight, 1% tolerance resistances probably should be used.
3. In the case of capacitors, percentage tolerances should parallel those given earlier for the resistors, for best results. As precision capacitors are relatively expensive, it may be desirable to use capacitors of higher tolerances, in which case trimming is required. In the case of low orders ($n \le 4$), 10% capacitors are quite often satisfactory.
4. The inverting gain of each stage of the filter is C_1/C_2. Gain adjustment can be made by trimming either C_1 or C_2.
5. For minimum dc offset, a resistance equal to R_2 can be placed in the noninverting input to ground.

3.6.3 BAND-PASS FILTERS

1. Standard resistance values of 5% tolerance normally yield acceptable results. In all cases, for best performance, resistance values close to those indicated should be used.
2. In the case of capacitors, 5% tolerances should be used for best results. As precision capacitors are relatively expensive, it may be desirable to use capacitors of higher tolerances, in which case trimming is generally required. In most cases, 10% capacitors are quite satisfactory.

PROBLEMS

3.1 Design an LP Butterworth filter 3 dB at 3 kHz and attenuation 40 db at 12 kHz with a gain of 10.

3.2 Design an LP Chebyshev 1 dB filter at 800 Hz and attenuation 35 dB at 2400 Hz with a gain of 1.

3.3 Design an LP Bessel filter of order six at 1 kHz with $K = 1$.

3.4 Design an HP Butterworth filter at 600 Hz and attenuation 40 dB at 150 Hz with a gain of 5.

3.5   Design an HP Chebyshev 3 dB filter at 350 Hz and attenuation 45 dB at 140 Hz and a gain of 9.

3.6   Design a Butterworth band-pass filter 3 dB from 300 to 3000 Hz and attenuation 30 dB below 60 Hz and above 15,000 Hz, with a gain of 10.

3.7   Design a Chebyshev 1-dB band-pass filter from 300 to 3000 Hz and attenuation 35 dB below 100 Hz and above 9000 Hz, with a gain of 1.

3.8   Design a narrow-band band-pass filter with $f_0 = 750$ Hz, $Q = 5$, and $K = 15$.

3.9   Design a narrow-band band-pass filter with two op-amps with $f_0 = 2.5$ kHz, $Q = 25$, and a gain of 10.

3.10 Design a Deliyannis filter with frequency of 500 Hz, $Q = 20$, and $K = 1$.

3.11 Design a wide-band band-reject Butterworth filter having 3 dB at 100 Hz and 400 Hz and greater than 40-dB attenuation between 350 and 114.3 Hz with a gain of 10.

3.12 Design a notch filter for $f_0 = 400$ Hz and $Q = 5$.

3.13 Design a notch filter for $f_0 = 450$ Hz, $Q = 8$, and $K = 10$.
       For the following filter:

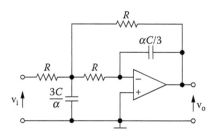

(a) Find the transfer function of the filter.
(b) Find the formulas to design the filter.
(c) From step (b) design a Butterworth filter 3 dB at 350 Hz and attenuation 50 dB at 1750 Hz.

3.14 For the following filter:

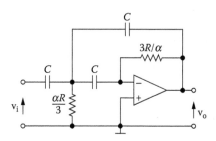

(a) Find the transfer function of the filter.
(b) Find the formulas to design the filter.
(c) From step (b) design a Butterworth filter 3 dB at 750 Hz and attenuation 50 dB at 150 Hz.

4 Filters with Three Op-Amps

In this chapter, a different type of multiple-feedback filter, called the state-variable filter and biquad filter, will be presented.

4.1 STATE-VARIABLE FILTER

The **state-variable** (SV) filter was initially developed for the analog computer. Although this type of filter uses at least three operational amplifiers, we nevertheless have the capability of simultaneous low-pass, high-pass, and band-pass output responses. When used with an additional op-amp, we can use the state-variable filter to form a notch filter.

Figure 4.1 shows that the state-variable filter, sometimes called a **universal filter**, is basically made up from a *summing amplifier*, two identical *integrators*, and a *damping network*. Because of the manner in which these functions are interconnected, we are able to simultaneously have the following filter responses:

1. A second-order low-pass filter
2. A second-order high-pass filter
3. A 1-pole band-pass filter

The cutoff frequency of the low-pass and high-pass response is **identical** to the center frequency of the band-pass response. In addition, the damping factor is equal to $1/Q$ for a band-pass filter, and is the same **for all three responses**.

Figure 4.2 shows the circuit connection for the state-variable filter. From this figure, we have:

$$V_{HP} = -\frac{R_2}{R_3}V_{LP} - \frac{R_2}{R_g}V_i + \frac{R_q}{R_q + R_1}\left(1 + \frac{R_2}{R_3} + \frac{R_2}{R_g}\right)V_{BP} \tag{4.1}$$

$$V_{BP} = -\frac{1}{sRC}V_{HP} \tag{4.2}$$

$$V_{LP} = -\frac{1}{sRC}V_{BP} \tag{4.3}$$

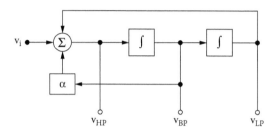

FIGURE 4.1 Basic state-variable filter.

4.1.1 LOW-PASS FILTER

From Equations (4.2) and (4.3), we have, respectively:

$$V_{HP} = -sRCV_{BP} \tag{4.4}$$

$$V_{BP} = -sRCV_{LH} \tag{4.5}$$

From Equations (4.4) and (4.5), we have:

$$V_{BP} = -sRCV_{LP} \tag{4.6}$$

From Equations (4.1), (4.5), and (4.6), we have:

$$s^2 R^2 C^2 V_{LP} = -\frac{R_2}{R_3} V_{LP} - \frac{R_2}{R_g} V_i - \frac{R_q}{R_q + R_1}\left(1 + \frac{R_2}{R_3} + \frac{R_2}{R_g}\right) sRCV_{LP} \qquad \therefore$$

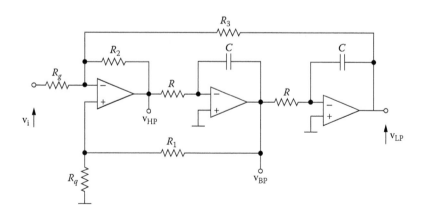

FIGURE 4.2 Inverting input state-variable filter.

$$\left[s^2 R^2 C^2 + \frac{R_q}{R_q + R_1}\left(1 + \frac{R_2}{R_3} + \frac{R_2}{R_g}\right) sRC + \frac{R_2}{R_3} \right] V_{LP} = -\frac{R_2}{R_g} V_i \qquad \therefore$$

$$H_{LP} = \frac{V_{LP}}{V_i} = -\frac{\dfrac{R_2}{R_g}}{s^2 R^2 C^2 + \dfrac{R_q}{R_q + R_1}\left(1 + \dfrac{R_2}{R_3} + \dfrac{R_2}{R_g}\right) sRC + \dfrac{R_2}{R_3}} \qquad (4.7)$$

$$H_{LP} = -\frac{K}{s^2 + as + b} \qquad (4.8)$$

where

$$K = \frac{R_3}{R_g} \quad (\text{for } s = 0) \qquad (4.9)$$

$$b = \omega_1^2 = \frac{R_2}{R_3}\frac{1}{R^2 C^2} \qquad (4.10)$$

$$\alpha = \frac{R_q}{R_q + R_1}\left(1 + \frac{R_2}{R_3} + \frac{R_2}{R_g}\right)\frac{1}{RC} \qquad (4.11)$$

4.1.1.1 Design Procedure

For the normalized filter, we have:

$$R = R_g = R_q = 1\,\Omega \qquad (4.12)$$

$$C = 1\,\text{F} \qquad (4.13)$$

From Equations (4.9) and (4.12), we have:

$$R_3 = K \qquad (4.14)$$

From Equations (4.10), (4.12), (4.13), and (4.14), we have:

$$b = \frac{R_2}{K} \qquad \therefore$$

$$R_2 = bK \qquad (4.15)$$

From Equations (4.11), (4.12), (4.13), (4.14), and (4.15), we have:

$$a = \frac{1}{1+R_1}\left(1+\frac{bK}{K}+\frac{bK}{1}\right)\frac{1}{1} \qquad \therefore$$

$$R_1 = \frac{1+b(1+K)}{\alpha} - 1 \qquad (4.16)$$

EXAMPLE 4.1

Design an LP state-variable Butterworth filter 3 dB at 1 kHz and attenuation 40 dB at 3.5 kHz with a gain of 1.

Solution

From Butterworth nomographs for $A_{max} = 3$ dB, $A_{min} = 40$ dB, and $f_s/f_1 = 3.5/1 = 3.5$, we find $n = 4$. Hence

$$R_n = R_{gn} = R_{qn} = 1 \ \Omega, \quad C_n = 1 \ F$$

$$R_{3n} = K = 1 \ \Omega$$

$$R_{2n} = bK = 1 \times 1 = 1 \ \Omega$$

First stage

$$\alpha = 1.848, \qquad b = 1.000$$

$$R_{1n} = \frac{1+(1+K)b}{a} - 1 = \frac{3}{a} - 1 = \frac{3}{1.848} - 1 = 0.623 \ \Omega$$

Second stage

$$\alpha = 0.765, \qquad b = 1.000$$

$$R_{1n} = \frac{3}{a} - 1 = \frac{3}{0.765} - 1 = 2.922 \ \Omega$$

Denormalization
We accept $ISF = 10^4$.

$$FSF = \frac{\omega_1}{\omega_n} = 2\pi \times 10^3$$

$$R = R_g = R_q = R_2 = R_3 = ISF \times R_n = 10^4 \times 1 \ \Omega = 10 \ k\Omega$$

$$C = \frac{C_n}{ISF \times FSF} = \frac{1}{2\pi \times 10^7} = 15.92 \ nF$$

First stage

$$R_1 = ISF \times R_{1n} = 10^4 \times 0.623 \ \Omega = 6.23 \ k\Omega$$

Second stage

$$R_1 = ISF \times R_{1n} = 10^4 \times 2.922 \ \Omega = 29.22 \ k\Omega$$

Figure 4.3a shows the designed filter, and its frequency response is shown in Figure 4.3b.

EXAMPLE 4.2

Design a low-pass state-variable Chebyshev 3-dB filter at 1 kHz and attenuation 35 dB at 2 kHz, with a gain of 5.

Solution

From Chebyshev 3-dB nomographs for $A_{max} = 3$ dB, $A_{min} = 35$ dB, and $f_s/f_1 = 2/1 = 2$, we find $n = 4$.

$$R_n = R_{gn} = R_{qn} = 1 \ \Omega, \quad C_n = 1 \ F$$

$$K_1 = K_2 = \sqrt{K} = \sqrt{5} = 2.236$$

$$R_{3n} = K_1 = 2.236$$

First stage

$$a = 0.411, \qquad b = 0.196$$

$$R_{2n} = bK = 0.196 \times 2.236 = 0.438 \ \Omega$$

$$R_{1n} = \frac{1 + (1 + K)b}{a} - 1 = \frac{1.634}{0.411} - 1 = 2.976 \ \Omega$$

Second stage

$$a = 0.170, \quad b = 0.903$$

$$R_{2n} = bK = 0.903 \times 2.236 \ \Omega = 2.019 \ \Omega$$

$$R_{1n} = \frac{1 + (1 + K)b}{a} - 1 = \frac{3.922}{0.170} - 1 = 22.071 \ \Omega$$

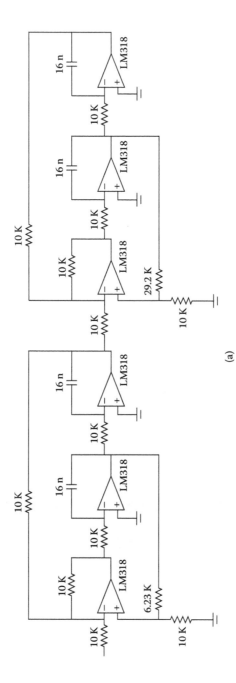

FIGURE 4.3 (a) S-V LP Butterworth filter, $f_1 = 1$ kHz, $K = 1$; (b) frequency response.

(a)

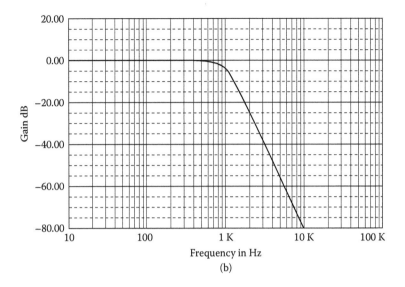

FIGURE 4.3 (Continued)

Denormalization

$$ISF = 10^4, \ FSF = \frac{\omega_1}{\omega_n} = 2\pi \times 10^3$$

$$R = R_g = R_q = ISF \times R_n = 10^4 \times 1 = 10 \ \text{k}\Omega$$

$$C = \frac{C_n}{ISF \times FSF} = \frac{1}{2\pi \times 10^7} = 15.92 \ \text{nF}$$

$$R_3 = ISF \times R_{3n} = 10^4 \times 2.236 \ \Omega = 22.36 \ \text{k}\Omega$$

First stage

$$R_1 = ISF \times R_{1n} = 10^4 \times 2.976 \ \Omega = 29.76 \ \text{k}\Omega$$

$$R_2 = ISF \times R_{2n} = 10^4 \times 0.438 \ \Omega = 4.38 \ \text{k}\Omega$$

Second stage

$$R_1 = ISF \times R_{1n} = 10^4 \times 22.071 \ \Omega = 220.71 \ \text{k}\Omega$$

$$R_2 = ISF \times R_{2n} = 10^4 \times 2.019 \ \Omega = 20.2 \ \text{k}\Omega$$

Figure 4.4a shows the designed filter, its frequency response is shown in Figure 4.4b, and its ripple in Figure 4.4c.

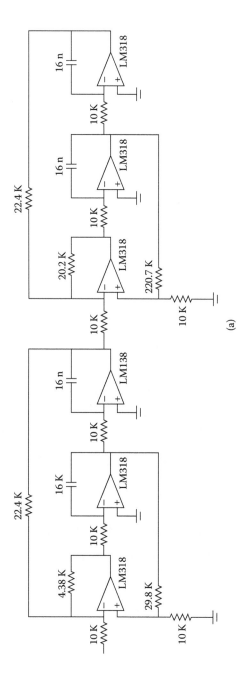

FIGURE 4.4 (a) S-V LP Chebyshev 3 dB, $f_1 = 1$ kHz, $K = 5$; (b) its frequency response; and (c) its ripple.

(a)

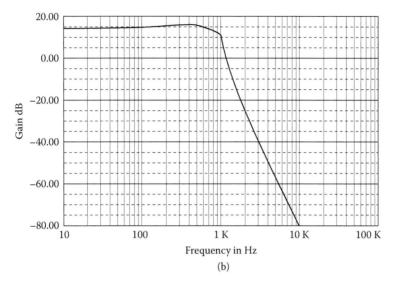

(b)

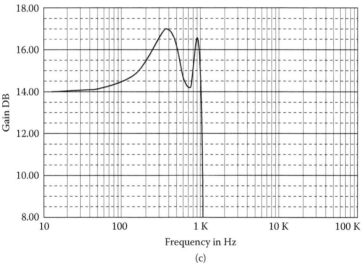

(c)

FIGURE 4.4 (Continued)

4.1.2 HIGH-PASS FILTER

From Equations (4.2) and (4.3), we have:

$$V_{LP} = -\frac{1}{sRC}\left(-\frac{1}{sRC}V_{HP}\right) = \frac{1}{s^2R^2C^2}V_{HP} \qquad (4.17)$$

and from Equations (4.1), (4.2), and (4.17):

$$V_{HP} = -\frac{R_2}{R_3} \cdot \frac{1}{s^2 R^2 C^2} V_{HP} - \frac{R_2}{R_g} V_i - \frac{R_q}{R_q + R_1}\left(1 + \frac{R_2}{R_3} + \frac{R_2}{R_g}\right)\frac{1}{sRC} V_{HP} \qquad \therefore$$

$$V_{HP}\left[1 + \frac{R_2}{R_3} \cdot \frac{1}{s^2 R^2 C^2} + \frac{R_q}{R_q + R_1}\left(1 + \frac{R_2}{R_3} + \frac{R_2}{R_g}\right)\frac{1}{sRC}\right] = -\frac{R_2}{R_g} V_i \qquad \therefore$$

$$V_{HP}\left[s^2 R^2 C^2 + \frac{R_q}{R_q + R_1}\left(1 + \frac{R_2}{R_3} + \frac{R_2}{R_g}\right)sRC + \frac{R_2}{R_3}\right] = -\frac{R_2}{R_g} s^2 R^2 C^2 V_i \qquad \therefore$$

$$H_{HP} = \frac{V_{HP}}{V_i} = -\frac{\dfrac{R_2}{R_g} s^2 R^2 C^2}{s^2 R^2 C^2 + \dfrac{R_q}{R_q + R_1}\left(1 + \dfrac{R_2}{R_3} + \dfrac{R_2}{R_g}\right)sRC + \dfrac{R_2}{R_3}} \qquad \therefore$$

$$H_{HP} = -\frac{\dfrac{R_2}{R_g} s^2}{s^2 + \dfrac{R_q}{R_q + R_1}\left(1 + \dfrac{R_2}{R_3} + \dfrac{R_2}{R_g}\right)\dfrac{1}{RC}s + \dfrac{R_2}{R_3}\dfrac{1}{R^2 C^2}} \qquad \therefore$$

$$H_{HP} = -\frac{Ks^2}{s^2 + as + b} \qquad (4.18)$$

where

$$b = \omega_2^2 = \frac{R_2}{R_3}\frac{1}{R^2 C^2} \qquad (4.19)$$

$$a = \frac{R_q}{R_q + R_1}\left(1 + \frac{R_2}{R_3} + \frac{R_2}{R_g}\right)\frac{1}{RC} \qquad (4.20)$$

For $s \to \infty$, we have:

$$K = H_{HP}(\infty) = \frac{R_2}{R_g} \qquad (4.21)$$

4.1.2.1 Design Procedure

For the normalized filter, we have:

$$R_n = R_{gn} = R_{qn} = 1 \ \Omega \qquad (4.22)$$

We can prove easily that

$$R_{2n} = K \tag{4.23}$$

$$R_{3n} = bK \tag{4.24}$$

$$R_{1n} = \frac{1+b(K+1)}{a} - 1 \tag{4.25}$$

EXAMPLE 4.3

Design an HP Butterworth 3-dB filter at 100 Hz and attenuation 40 dB at 28.6 Hz, with a gain of 1.

Solution

From Butterworth nomographs for A_{max} = 3 dB, A_{min} = 40 dB, and f_2/f_s = 100/28.6 @ 3.5, we find $n = 4$.

$$R_{qn} = R_{gn} = R_n = 1 \ \Omega$$

$$C_n = 1 \ F$$

$$R_{2n} = \sqrt{K} = 1 \ \Omega$$

First stage

$$a = 1.848, \qquad b = 1.000$$

$$R_{1n} = \frac{1+b(K+1)}{a} - 1 = \frac{1+2}{1.848} - 1 = 0.623 \ \Omega$$

$$R_{3n} = bK = 1 \ \Omega$$

Second stage

$$a = 0.765, \qquad b = 1.000$$

$$R_{1n} = \frac{1+b(K+1)}{a} - 1 = \frac{3}{0.765} - 1 = 2.922 \ \Omega$$

$$R_{3n} = bK = 1 \ \Omega$$

Denormalization

$$ISF = 10^4, \ FSF = \frac{\omega_2}{\omega_n} = 2\pi f_2 = 2\pi \times 100 = 200\pi$$

$$C = \frac{C_n}{ISF \times FSF} = \frac{1}{2\pi \times 10^6} = 0.159 \ \mu F$$

$$R_2 = ISF \times R_{2n} = 10 \times 1 = 10 \ k\Omega$$

First stage

$$R_1 = ISF \times R_{1n} = 10^4 \times 0.623 \ \Omega = 6.23 \ k\Omega$$

$$R_3 = ISF \times R_{3n} = 10^4 \times 1 \ \Omega = 10 \ k\Omega$$

Second stage

$$R_1 = ISF \times R_{1n} = 10^4 \times 2.922 \ \Omega = 29.22 \ k\Omega$$

$$R_3 = ISF \times R_{3n} = 10_4 \times 1 \ \Omega = 10 \ k\Omega$$

Figure 4.5a shows the designed filter, and its frequency response is shown in Figure 4.5b.

EXAMPLE 4.4

Design an HP Chebyshev 3-dB filter at 100 Hz and attenuation 40 dB at 40 Hz, with a gain of 5.

Solution

From Chebyshev nomographs for $A_{max} = 3$ dB, $A_{min} = 40$ dB, and $f_2/f_s = 100/40 = 25$, we find $n = 4$.

$$K = \sqrt{5}$$

$$R_n = R_{gn} = R_{qn} = 1 \ \Omega$$

$$C_n = 1 \ F$$

First stage

$$a = 0.411, \quad b = 0.196$$

$$R_{1n} = \frac{1 + b(1 + K)}{a} - 1 = \frac{1 + 0.196 \times 3.236}{0.411} - 1 = 2.976 \ \Omega$$

$$R_{2n} = K = 2.236 \ \Omega$$

$$R_{3n} = bK = 0.196 \times 2.236 = 0.438 \ \Omega$$

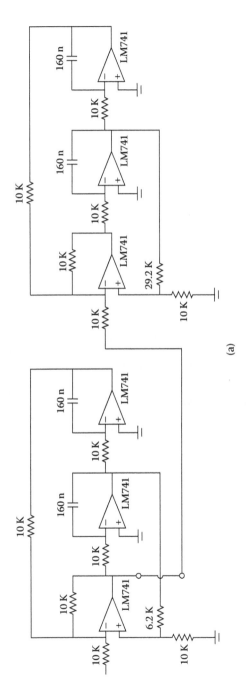

(a)

FIGURE 4.5 (a) Fourth-order S-V HP Butterworth 3 dB, $f_2 = 100$ Hz, $K = 1$; (b) frequency response.

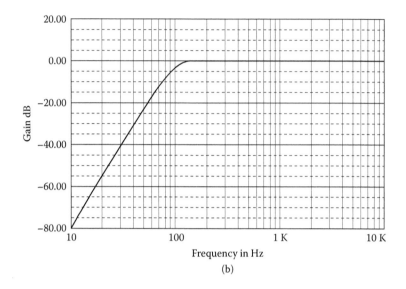

FIGURE 4.5 (Continued)

Second stage

$$a = 0.170, \qquad b = 0.903$$

$$R_{1n} = \frac{1 + 0.903 \times 3.236}{0.170} - 1 = 22.071 \ \Omega$$

$$R_{2n} = 2.236 \ \Omega$$

$$R_{3n} = 0.903 \times 2.236 = 2.019 \ \Omega$$

Denormalization

$$ISF = 10^4, \ FSF = \frac{\omega_2}{\omega_n} = 2\pi f_2 = 2\pi \times 100$$

$$R = R_g = R_q = 10 \ \text{k}\Omega$$

$$C = \frac{C_n}{ISF \times FSF} = \frac{1}{2\pi \times 10^6} = 0.16 \ \mu\text{F}$$

$$R_2 = ISF \times R_{2n} = 10^4 \times 2.236 \ \Omega = 22.4 \ \text{k}\Omega$$

First stage

$$R_1 = ISF \times R_{1n} = 10^4 \times 2.976 \ \Omega = 29.8 \ \text{k}\Omega$$

$$R_3 = ISF \times R_{3n} = 10^4 \times 0.438 \ \Omega = 4.38 \ \text{k}\Omega$$

Second stage

$$R_1 = ISF \times R_{1n} = 10^4 \times 22.071 \ \Omega = 220.7 \ k\Omega$$

$$R_3 = ISF \times R_{3n} = 10^4 \times 2.019 \ \Omega = 20.2 \ k\Omega$$

Figure 4.6a shows the designed filter, and its frequency response is shown in Figure 4.6b.

EXAMPLE **4.5**

Design a Butterworth band-pass filter 3 dB from 300 to 3000 Hz and attenuation 30 dB below 50 Hz and above 18,000 Hz, with gain 1.

Solution

(a) *Low-pass filter*
From Butterworth nomographs for $A_{max} = 3$ dB, $A_{min} = 30$ dB, and $f_s/f_1 = 18,000/3000 = 6$, we find $n = 2$. Hence

$$a = 1.414, \qquad b = 1.000$$

$$R_n = R_{gn} = R_{qn} = R_{3n} = 1 \ \Omega$$

$$R_{2n} = bK = 1 \times 1 = 1 \ \Omega$$

$$R_{1n} = \frac{1+b+bK}{a} - 1 = \frac{3}{a} - 1 = \frac{3}{1.414} - 1 = 1.122 \ \Omega$$

Denormalization

$$ISF = 10^4, \ FSF = \frac{\omega_1}{\omega_n} = 2\pi \ f_1 = 2\pi \times 3000 = 6\pi \times 10^3$$

$$R = R_g = R_q = R_2 = R_3 = ISF \times R_n = 10^4 \times 1 \ \Omega = 10 \ k\Omega$$

$$R_1 = ISF \times R_{1n} = 10^4 \times 1.122 \ \Omega = 11.2 \ k\Omega$$

$$C = \frac{C_n}{ISF \times FSF} = \frac{1}{6\pi \times 10^7} = 5.3 \ nF$$

(b) *High-pass filter*
From Butterworth nomographs for $A_{max} = 3$ dB, $A_{min} = 30$ dB, and $f_2/f_s = 300/50 = 6$, we find $n = 2$.

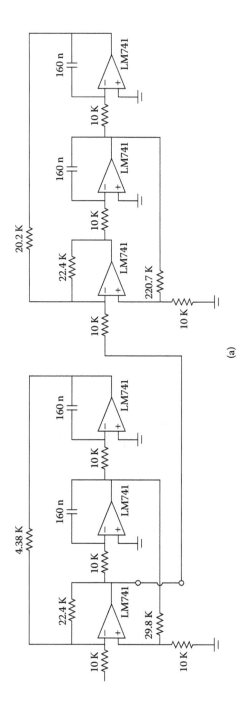

(a)

FIGURE 4.6 (a) Fourth-order S–V HP Chebyshev 3 dB, $f_2 = 100$ Hz, $K = 5$; (b) frequency response.

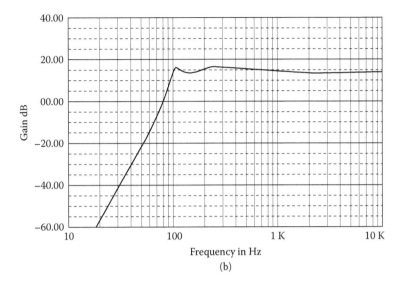

FIGURE 4.6 (Continued)

Denormalization

$$ISF = 10^4, \quad FSF = \frac{\omega_2}{\omega_n} = \frac{2\pi f_2}{1} = 2\pi \times 300 = 6\pi \times 10^2$$

$$C = \frac{C_n}{ISF \times FSF} = \frac{1}{6\pi \times 10^2} = 53.1 \ \text{nF}$$

Figure 4.7a shows the designed filter, and its frequency response is shown in Figure 4.7b.

EXAMPLE 4.6

Design a wide-band band-reject Butterworth filter having 3 dB at 100 and 1000 Hz and greater than 20 dB of attenuation between 500 and 360 Hz with a gain of 5.

Solution

Because the ratio of upper cutoff to lower cutoff is well in excess of an octave, a wide-band approach can be used. First, separate the specification into individual low-pass and high-pass requirements.

(a) *Low-pass filter*

From Butterworth nomographs for $A_{max} = 3$ dB, $A_{min} = 20$ dB, and $f_s/f_1 = 500/100 = 5$, we find $n = 2$.

$$a = 1.414, \quad b = 1.000$$

$$R_n = R_{gn} = R_{qn} = 1 \ \Omega, \quad C_n = 1 \ \text{F}$$

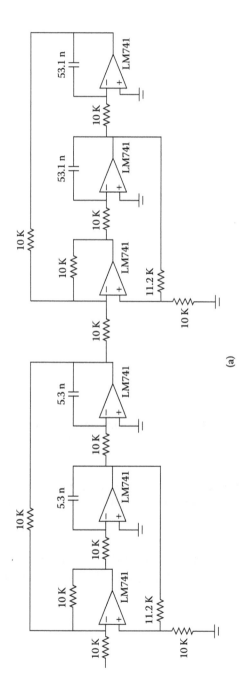

(a)

FIGURE 4.7 (a) Second-order S-V Butterworth BP 300 to 3000 Hz, $K = 1$; (b) frequency response.

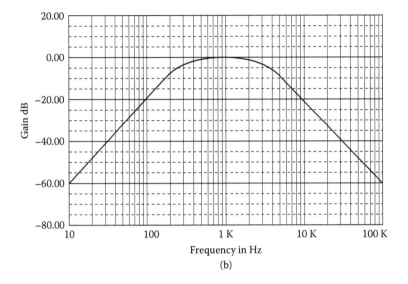

FIGURE 4.7 (Continued)

For $K = 1$, we have:

$$R_{3n} = K = 1 \ \Omega, \quad R_{2n} = bK = 1 \ \Omega$$

$$R_{1n} = \frac{1+b+bK}{a} - 1 = \frac{3}{a} - 1 = \frac{3}{1.414} - 1 = 1.122 \ \Omega$$

Denormalization

$$ISF = 10^4, \ FSF = \frac{\omega_1}{\omega_n} = 2\pi f_1 = 2\pi \times 100$$

$$C = \frac{C_n}{ISF \times FSF} = \frac{1}{2\pi \times 10^6} = 0.159 \ \mu F$$

$$R = R_g = R_q = R_3 = R_2 = ISF \times R_n = 10^4 \times 1 \ \Omega = 10 \ k\Omega$$

$$R_1 = ISF \times R_{1n} = 10^4 \times 1.122 \ \Omega = 11.22 \ k\Omega$$

(b) *High-pass filter*

Denormalization

$$ISF = 10^4, \ FSF = \frac{\omega_2}{\omega_n} = 2\pi f_2 = 2\pi \times 10^3$$

$$C = \frac{C_n}{ISF \times FSF} = \frac{1}{2\pi \times 10^7} = 15.9 \ nF$$

Figure 4.8a shows the designed filter, and its frequency response is shown in Figure 4.8b.

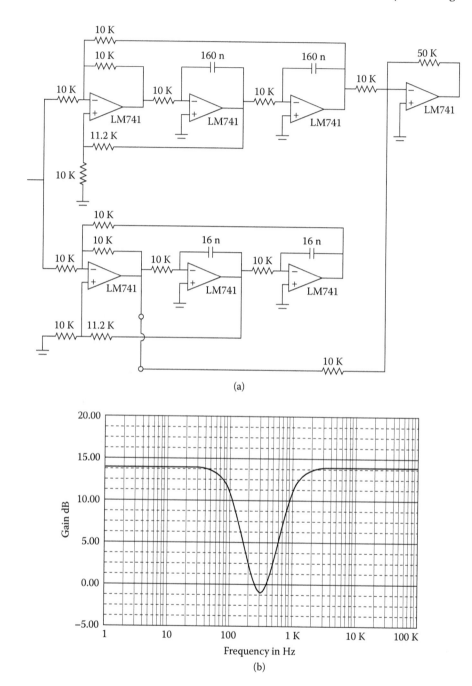

(a)

(b)

FIGURE 4.8 (a) Wide-band band-reject S-V filter, 100 to 1000 Hz, $K = 5$; (b) frequency response.

4.1.3 NARROW-BAND BAND-PASS FILTER

From Equations (4.1), (4.2), and (4.3), we have:

$$-sRCV_{BP} = -\frac{R_2}{R_3}\left(-\frac{1}{sRC}V_{BP}\right) - \frac{R_2}{R_g}V_i - \frac{R_q}{R_q+R_1}\left(1+\frac{R_2}{R_3}+\frac{R_2}{R_g}\right)V_{BP} \quad \therefore$$

$$\left[sRC+\frac{R_2}{R_3}\frac{1}{sRC}+\frac{R_q}{R_q+R_1}\left(1+\frac{R_2}{R_3}+\frac{R_2}{R_g}\right)\right]V_{BP} = \frac{R_2}{R_g}V_i \quad \therefore$$

$$\left[s^2R^2C^2+\frac{R_q}{R_q+R_1}\left(1+\frac{R_2}{R_3}+\frac{R_2}{R_g}\right)sRC+\frac{R_2}{R_3}\right]V_{BP} = \frac{R_2}{R_g}sRCV_i \quad \therefore$$

$$H_{BP}(s) = \frac{V_{BP}}{V_i} = \frac{\dfrac{R_2}{R_g}sRC}{s^2R^2C^2+\dfrac{R_q}{R_q+R_1}\left(1+\dfrac{R_2}{R_3}+\dfrac{R_2}{R_g}\right)sRC+\dfrac{R_2}{R_3}} \quad \therefore$$

$$H_{BP} = \frac{\dfrac{R_2}{R_g}sRC}{R^2C^2\left[s^2+\dfrac{R_q}{R_q+R_1}\left(1+\dfrac{R_2}{R_3}+\dfrac{R_2}{R_g}\right)\dfrac{1}{RC}s+\dfrac{R_2}{R_3}\dfrac{1}{R^2C^2}\right]} \quad \therefore$$

$$H_{BP} = \frac{\dfrac{R_2}{R_g}\dfrac{1}{RC}s}{s^2+\dfrac{R_q}{R_q+R_1}\left(1+\dfrac{R_2}{R_3}+\dfrac{R_2}{R_g}\right)\dfrac{1}{RC}s+\dfrac{R_2}{R_3}\dfrac{1}{R^2C^2}} \quad (4.26)$$

If we put

$$\omega_0^2 = \frac{R_2}{R_3}\frac{1}{R^2C^2} \quad \therefore$$

$$\omega_0 = \frac{\sqrt{\dfrac{R_2}{R_3}}}{RC} \quad (4.27)$$

Hence

$$H_{BP} = \frac{\dfrac{R_2}{R_g}\dfrac{1}{RC}s}{s^2 + \dfrac{R_q}{R_q + R_1}\left(1 + \dfrac{R_2}{R_3} + \dfrac{R_2}{R_g}\right)\dfrac{1}{RC}s + \omega_0^2} \qquad \therefore$$

$$H_{BP} = \frac{\dfrac{R_2}{R_g}\dfrac{1}{\omega_0 RC}\left(\dfrac{s}{\omega_0}\right)}{\left(\dfrac{s}{\omega_0}\right)^2 + \dfrac{R_q}{R_q + R_1}\left(1 + \dfrac{R_2}{R_3} + \dfrac{R_2}{R_g}\right)\dfrac{1}{\omega_0 RC}\left(\dfrac{s}{\omega_o}\right) + 1}$$

If we put

$$a = \frac{R_q}{R_q + R_1}\left(1 + \frac{R_2}{R_3} + \frac{R_2}{R_g}\right)\frac{1}{\omega_0 RC} \qquad (4.28)$$

we have

$$H_{BP} = \frac{\dfrac{R_2}{R_g}\dfrac{1}{\omega_0 RC}\left(\dfrac{s}{\omega_0}\right)}{\left(\dfrac{s}{\omega_0}\right)^2 + a\left(\dfrac{s}{\omega_0}\right) + 1} \qquad (4.29)$$

For $s = j\omega_o$, from Equations (2.27) and (4.26), we have:

$$K = \frac{\dfrac{R_2}{R_g}\dfrac{1}{\omega_0 RC}}{\dfrac{R_q}{R_q + R_1}\left(1 + \dfrac{R_2}{R_3} + \dfrac{R_2}{R_g}\right)\dfrac{1}{\omega_o RC}} \qquad \therefore$$

$$K = \frac{\dfrac{R_2}{R_g}}{\dfrac{R_q}{R_q + R_1}\left(1 + \dfrac{R_2}{R_3} + \dfrac{R_2}{R_g}\right)} \qquad (4.30)$$

4.1.3.1 Design Procedure

For the normalized filter, $\omega_0 = 1$ rad/s and

$$R = R_g = R_2 = 1 \ \Omega \tag{4.31}$$

$$C = 1 \ F \tag{4.32}$$

From Equations (4.25), (4.26), and (4.28), we have, respectively:

$$R_3 = 1 \ \Omega \tag{4.33}$$

$$\alpha = \frac{3R_q}{R_q + R_1} \qquad \therefore$$

$$R_q + R_1 = \frac{3R_q}{a} = 3QR_q \qquad \therefore$$

$$R_1 = (3Q - 1)R_q \tag{4.34}$$

From Equations (4.28) and (4.32), we have:

$$K = \frac{1}{\dfrac{3R_q}{R_q + (3Q - 1)R_q}} = \frac{1 + 3Q - 1}{3} \qquad \therefore$$

$$K = Q \tag{4.35}$$

EXAMPLE 4.7

Design a narrow-band band-pass filter with a center frequency of 1 kHz and $Q = 40$.

Solution

$$R_n = R_{gn} = R_{qn} = R_{2n} = R_{3n} = 1 \ \Omega$$

$$C_n = 1 \ F$$

$$R_{1n} = (3Q - 1)R_{qn} = 119R_{qn} = 119 \ \Omega$$

$$K = Q = 40 \quad \text{or} \quad 32 \ dB$$

Denormalization

$$ISF = 10^3, \ FSF = \frac{\omega_0}{\omega_n} = 2\pi f_0 = 2\pi \times 10^3 \qquad \therefore$$

$$R = R_g = R_q = R_2 = R_3 = ISF \times R_n = 10^3 \times 1 \ \Omega = 10 \ k\Omega$$

$$R_1 = ISF \times R_{1n} = 10^4 \times 119 \ \Omega = 119 \ k\Omega$$

$$C = \frac{C_n}{ISF \times FSF} = \frac{1}{2\pi \times 10^6} = 0.159 \ \mu F$$

Figure 4.9a shows the designed filter, and its frequency response is shown in Figure 4.9b.

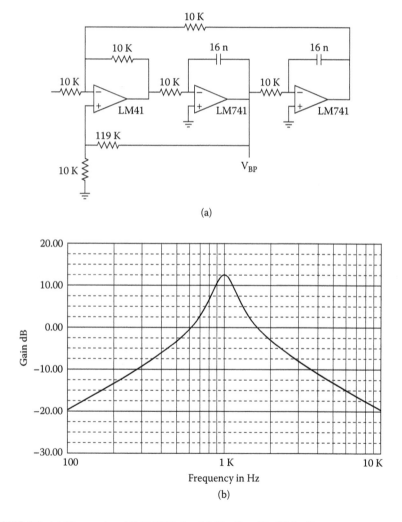

(a)

(b)

FIGURE 4.9 (a) Narrow-band S-V BPF, $f_0 = 1$ kHz, $Q = 40$; (b) its frequency response.

A number of manufacturers offer "ready-to-go" state-variable filters. Depending on whether the unity-gain or four op-amp state-variable types are required, only three or four external resistors are required to "program" the filter to your requirements.

The commercial devices essentially follow the aforementioned design, except that the frequency-determining capacitors and the resistors associated with the summing amplifier are already inside the filter. An additional op-amp section, which is uncommitted, can be used to form notch filters. Some of the commercial types available are listed as follows:

AF100 Universal Active Filter, National Semiconductor
UAF41 state-variable filter, Burr-Brown
ACF7092, 16-pin DIP, General Instrument Corporation
FS-60, Active Filter, Kinetic Technology Corporation

EXAMPLE 4.8

Design a 100-Hz notch filter with $Q = 25$.

Solution

$$R_n = R_{gn} = R_{qn} = R_{2n} = R_{3n} = 1 \ \Omega$$

$$C_n = 1 \ F$$

$$R_{1n} = 3Q - 1 = 3 \times 25 - 1 = 74 \ \Omega$$

$$K = Q = 25 \quad \text{or} \quad 28 \ dB$$

Denormalization

$$ISF = 10^4, \quad FSF = \frac{\omega_0}{\omega_n} = 2\pi \times 10^2$$

$$R = R_g = R_q = R_2 = R_3 = ISF \times R_n = 10^4 \times 1 \ \Omega = 10 \ k\Omega$$

$$R_1 = ISF \times R_{1n} = 10^4 \times 74 \ \Omega = 740 \ k\Omega$$

$$C = \frac{C_n}{ISF \times FSF} = \frac{1}{2\pi \times 10^6} = 0.159 \ \mu F$$

To complete the design, add the two-input op-amp summing amplifier to the LP and HP outputs, giving the final circuit of Figure 4.10a; its frequency response is shown in Figure 4.10b.

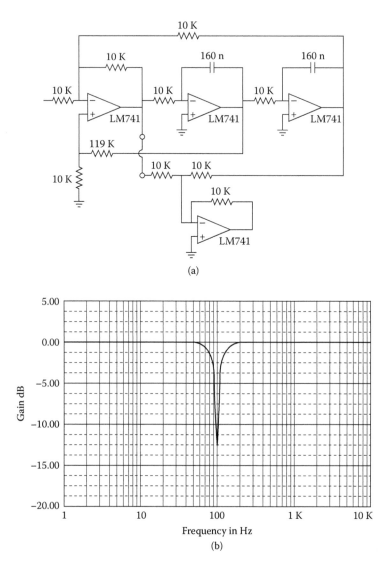

FIGURE 4.10 (a) State-variable notch filter, $f_0 = 100$ Hz, $Q = 25$; (b) its frequency response.

4.2 BIQUAD FILTERS

The biquad filter demonstrates a particular useful characteristic, **constant bandwidth**. The band-pass output of the state-variable filter has a Q that is constant as the center frequency is varied. This in turn implies that the bandwidth narrows at lower and at higher frequencies. The biquad filter, although appearing very similar to the state-variable filter, has a bandwidth that is fixed regardless of center frequency. This type of filter (shown in Figure 4.11) is useful in applications such as **spectrum analyzers**, which require a filter with a fixed bandwidth. This characteristic is also needed in **telephone applications**, in which a group of identical absolute bandwidth channels

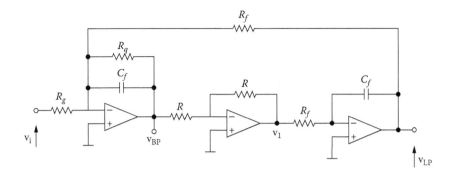

FIGURE 4.11 Biquad filter.

is needed at different center frequencies. The center frequency is easily tuned by merely adjusting the value of R_f. Also, Q may be adjusted by changing the value of R_q, and the gain of the filter may be changed by adjusting the value of R_g. The biquad filter is capable of attaining high values of Q, in the neighborhood of 100, and is a much more stable network than those discussed in the previous band-pass filters.

From this figure, we have:

$$V_{BP} = -\frac{Z}{R_g}V_i - \frac{Z}{R_f}V_{LP} \qquad (4.36)$$

$$V_1 = -V_B \qquad (4.37)$$

$$V_{LP} = -\frac{1}{sR_fC_f}V_1 \qquad (4.38)$$

$$Y = \frac{1}{Z} = \frac{1}{R_q} + sC_f = \frac{1 + sR_qC_f}{R_q} \qquad (4.39)$$

4.2.1 Narrow-Band Band-Pass Filter

From Equations (4.37) and (4.38), we have:

$$V_{LP} = -\frac{1}{sR_fC_f}(-V_{BP}) = \frac{1}{sR_fC_f}V_{BP} \qquad (4.40)$$

From Equations (4.36) and (4.40), we have:

$$V_{BP} = -\frac{Z}{R_g}V_i - \frac{Z}{R_f}\cdot\frac{1}{sR_fC_f}V_{BP} \qquad \therefore$$

$$V_{BP}\left(1 + \frac{Z}{sR_f^2C_f}\right) = -\frac{Z}{R_g}V_i \qquad \therefore$$

$$V_{BP} \frac{1+sR_f^2C_f Z}{sR_f^2C_f} = -\frac{Z}{R_g}V_i \qquad \therefore$$

$$H_{BP} = \frac{V_{BP}}{V_i} = -\frac{sR_f^2C_f Z}{R_g(sR_f^2C_f + Z)} = -\frac{sR_f^2C_f}{R_g\left(1+\dfrac{sR_f^2C_f}{Z}\right)} \qquad \therefore$$

$$H_{BP} = -\frac{\dfrac{sR_f^2C_f}{R_g}}{1+sR_f^2C_f\dfrac{1+sR_qC_f}{R_q}} \qquad \therefore$$

$$H_{BP} = -\frac{\dfrac{sR_f^2C_f R_q}{R_g}}{R_q + sR_f^2C_f(1+sR_qC_f)} \qquad \therefore$$

$$H_{BP} = -\frac{\dfrac{sR_f^2C_f R_q}{R_g}}{s^2 R_f^2 R_q C_f^2 + sR_f^2C_f + R_q} \qquad \therefore$$

$$H_{BP} = -\frac{\dfrac{1}{R_gC_f}s}{s^2 + \dfrac{1}{R_qC_f}s + \omega_o^2}$$

where

$$\omega_0 = \frac{1}{R_fC_f} \qquad\qquad (4.41)$$

\

$$H_{BP} = -\frac{\dfrac{1}{\omega_0 R_gC_f}\left(\dfrac{s}{\omega_0}\right)}{\left(\dfrac{s}{\omega_0}\right)^2 + \dfrac{1}{Q}\left(\dfrac{s}{\omega_0}\right) + 1} \qquad\qquad (4.42)$$

where

$$Q = \omega_0 R_qC_f \qquad\qquad (4.43)$$

For $s = j\omega_0$, Equation (4.42) becomes:

$$K = H_{BP}(\omega_0) = -\frac{\dfrac{1}{\omega_0 R_g C_f}}{\dfrac{1}{\omega_0 R_q C_f}} \qquad \therefore$$

$$K = -\frac{R_q}{R_g} \tag{4.44}$$

4.2.1.1 Design Procedure

From Equation (4.41), we have:

$$R_f = \frac{1}{\omega_0 C_f} \tag{4-45}$$

From Equations (4.43) and (4.45), we have:

$$R_q = \frac{Q}{\omega_0 C_f} = QR_f \tag{4.46}$$

and from Equation (4.44)

$$R_g = \frac{R_q}{K} \tag{4.47}$$

From Equation (7.46), we have:

$$Q = \frac{R_q}{R_f} \tag{4.48}$$

But

$$BW = \frac{\omega_0}{Q}$$

Upon close examination of Q, it should be apparent that as R_f increases [lower center frequency, Equation (4.45)], Q drops, keeping the bandwidth constant as predicted. This type of tuning is satisfactory over two to three decades.

For the normalized filter,

$$\omega_0 = 1 \text{ rad/s}$$

and

$$C_f = 1 \text{ F} \qquad\qquad (4.49)$$

and from Equation (4.45), we have:

$$R_f = 1 \ \Omega \qquad\qquad (4.50)$$

From Equation (4.46), we have:

$$R_q = Q \qquad\qquad (4.51)$$

and from Equations (4.48) and (4.52)

$$R_g = \frac{Q}{K} \qquad\qquad (4.52)$$

EXAMPLE 4.9

Design a narrow-band band-pass filter with $f_0 = 1$ kHz, $Q = 50$, and a gain of 5.

Solution

$$R_n = R_{fn} = 1 \ \Omega$$

$$C_n = 1 \text{ F}$$

$$R_{qn} = Q = 50 \ \Omega$$

$$R_{gn} = \frac{R_{qn}}{K} = \frac{Q}{K} = \frac{50}{5} = 10 \ \Omega$$

Denormalization

$$ISF = 10^4, \ FSF = \frac{\omega_0}{\omega_n} = 2\pi f_0 = 2\pi \times 10^3$$

$$R = R_f = ISF \times R_n = 10^4 \times 1 \ \Omega = 10 \text{ k}\Omega$$

$$R_g = ISF \times R_{gn} = 10^4 \times 10 \ \Omega = 100 \ \text{ k}\Omega$$

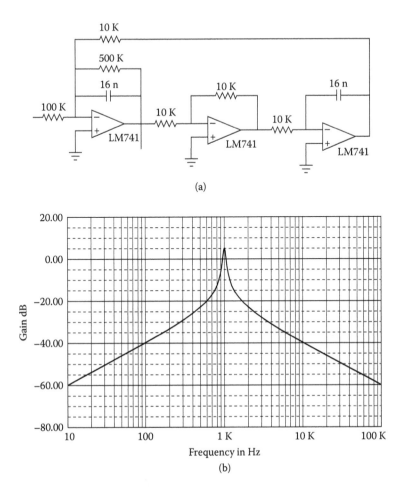

(a)

(b)

FIGURE 4.12 (a) Narrow-band band-pass filter, $f_0 = 1$ kHz, $Q = 50$, $K = 5$; (b) its frequency response.

$$R_q = ISF \times R_{qn} = 10^4 \times 50 \ \Omega = 500 \ \text{k}\Omega$$

$$C = \frac{C_n}{ISF \times FSF} = \frac{1}{2 \ \times 10^7} = 15.92 \ \text{nF}$$

Figure 4.12a shows the designed filter, and its frequency response is shown in Figure 4.12b.

4.2.2 Low-Pass Filter

From Equation (4.40), we have:

$$V_{BP} = sR_f C_f V_{LP} \tag{4.53}$$

From Equations (4.36) and (4.53), we have:

$$sR_fC_fV_{LP} = -\frac{Z}{R_g}V_i - \frac{Z}{R_f}V_{LP} \qquad \therefore$$

$$\left(sR_fC_f + \frac{Z}{R_f}\right)V_{LP} = -\frac{Z}{R_g} \qquad \therefore$$

$$\left[sR_fC_f + \frac{R_q}{R_f(1+sR_qC_f)}\right]V_{LP} = -\frac{R_q}{R_g(1+sR_qC_f)}V_i \qquad \therefore$$

$$\left(sR_fC_f(1+sR_qC_f) + \frac{R_q}{R_f}\right)V_{LP} = -\frac{R_q}{R_g}V_i \qquad \therefore$$

$$\left(s^2R_fR_qC_f^2 + sR_fC_f + \frac{R_q}{R_f}\right)V_{LP} = -\frac{R_q}{R_g}V_i \qquad \therefore$$

$$H_{LP} = \frac{V_{LP}}{V_i} = -\frac{\dfrac{R_q}{R_g}}{s^2R_fR_qC_f^2 + sR_fC_f + \dfrac{R_q}{R_f}} \qquad \therefore$$

$$H_{LP} = -\frac{\dfrac{1}{R_fR_qC_f^2}}{s^2 + \dfrac{1}{R_qC_f}s + \dfrac{1}{R_f^2C_f^2}}$$

$$H_{LP} = -\frac{\dfrac{1}{R_fR_gC_f^2}}{s^2 + as + b} \tag{4.54}$$

where

$$b = \omega_1^2 = \frac{1}{R_f^2C_f^2} \tag{4.55}$$

and

$$\alpha = \frac{1}{R_qC_f} \tag{4.56}$$

For $s = j\omega = 0$, from Equation (4.54), we have:

$$H_{LP} = K = -\frac{R_f}{R_g} \tag{4.57}$$

4.2.2.1 Design Procedure

From Equation (4.55), we have:

$$R_f = \frac{1}{C_f\sqrt{b}} \tag{4.58}$$

From Equation (4.56), we have:

$$R_q = \frac{1}{aC_f} \tag{4.59}$$

and from Equation (4.57)

$$R_g = \frac{R_f}{K} \tag{4.60}$$

For the normalized filter, $\omega_1 = 1$ rad/s and

$$C_{fn} = 1 \text{ F} \tag{4.61}$$

we have:

$$R_{fn} = \frac{1}{\sqrt{b}} \tag{4.62}$$

$$R_{qn} = \frac{1}{a} \tag{4.63}$$

$$R_{qn} = \frac{R_{fn}}{K} = \frac{1}{K\sqrt{b}} \tag{4.64}$$

EXAMPLE 4.10

Design a second-order low-pass Butterworth filter with a cutoff frequency of 1 kHz and a gain of 10.

Solution

$$a = 1.414, \qquad b = 1.009$$

$$R_{qn} = \frac{1}{a} = \frac{1}{1.414} = 0.708 \ \Omega$$

$$R_{gn} = \frac{1}{K\sqrt{b}} = \frac{1}{10 \times 1} = 0.1 \ \Omega$$

$$R_{fn} = \frac{1}{\sqrt{b}} = \frac{1}{1} = 1 \ \Omega$$

Denormalization

$$ISF = 10^4, \ FSF = \frac{\omega_1}{\omega_n} = 2\pi \, f_1 = 2\pi \times 10^3 \qquad \therefore$$

$$C_f = \frac{C_{fn}}{ISF \times FSF} = \frac{1}{2\pi \times 10^7} = 15.92 \ \text{nF}$$

$$R_q = ISF \times R_{qn} = 10^4 \times 0.707 \ \Omega \cong 7.1 \ \text{k}\Omega$$

$$R_g = ISF \times R_{gn} = 10^4 \times 0.1 \ \Omega = 1 \ \text{k}\Omega$$

$$R_f = ISF \times R_{fn} = 10^4 \times 1 \ \Omega = 10 \ \text{k}\Omega$$

$$R = ISF \times R_n = 10^4 \times 1 \ \Omega = 10 \ \text{k}\Omega$$

Figure 4.13a shows the designed filter, and its frequency response is shown in Figure 4.13b.

EXAMPLE 4.10

Design an LP Chebyshev 3-dB filter at 1 kHz and attenuation 40 dB at 2.5 kHz, with a gain of 10.

Solution

From Chebyshev nomographs for $A_{max} = 3$ dB, $A_{min} = 40$ dB, and $f_s/f_1 = 2.5/1 = 2.5$, we find $n = 4$.

$$K = \sqrt{10} = 3.162$$

$$R_n = 1 \ \Omega, \qquad C_{fn} = 1 \ \text{F}$$

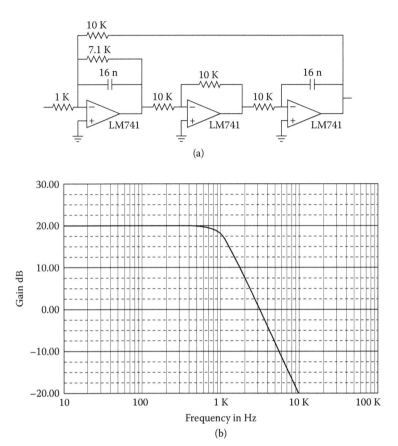

(a)

(b)

FIGURE 4.13 (a) Low-pass Butterworth filter, $f_1 = 1$ kHz, $K = 10$; (b) frequency response.

First stage

$$a = 0.411, \qquad b = 0.196$$

$$R_{qn} = \frac{1}{a} = \frac{1}{0.411} = 2.433 \ \Omega$$

$$R_{gn} = \frac{1}{K\sqrt{b}} = \frac{1}{3.162 \times \sqrt{0.196}} = 0.714 \ \Omega$$

$$R_{fn} = \frac{1}{\sqrt{b}} = \frac{1}{\sqrt{0.196}} = 2.259 \ \Omega$$

Second stage

$$a = 0.170, \quad b = 0.903$$

$$R_{qn} = \frac{1}{a} = \frac{1}{0.170} = 5.882 \ \Omega$$

$$R_{gn} = \frac{1}{K\sqrt{b}} = \frac{1}{3.162 \times \sqrt{0.903}} = 0.333 \ \Omega$$

$$R_{fn} = \frac{1}{\sqrt{b}} = \frac{1}{\sqrt{0.903}} = 1.052 \ \Omega$$

Denormalization

$$ISF = 10^4, \ FSF = \frac{\omega_1}{\omega_n} = 2\pi f_1 = 2\pi \times 10^3$$

$$R = ISF \times R_n = 10^4 \times 1 \ \Omega = 10 \ \text{k}\Omega$$

$$C_f = \frac{C_{fn}}{ISF \times FSF} = \frac{1}{2\pi \times 10^7} = 15.9 \ \text{nF}$$

First stage

$$R_q = ISF \times R_{qn} = 10^4 \times 2.433 \ \Omega = 24.3 \ \text{k}\Omega$$

$$R_g = ISF \times R_{gn} = 10^4 \times 0.714 \ \Omega = 7.1 \ \text{k}\Omega$$

$$R_f = ISF \times R_{fn} = 10^4 \times 2.26 \ \Omega = 22.6 \ \text{k}\Omega$$

Second stage

$$R_q = ISF \times R_{qn} = 10^4 \times 5.882 \ \Omega = 58.8 \ \text{k}\Omega$$

$$R_g = ISF \times R_{gn} = 10^4 \times 0.333 \ \Omega = 3.3 \ \text{k}\Omega$$

$$R_f = ISF \times R_{fn} = 10^4 \times 1.052 \ \Omega = 10.5 \ \text{k}\Omega$$

Figure 4.14a shows the designed filter, and its frequency response is shown in Figure 4.14b.

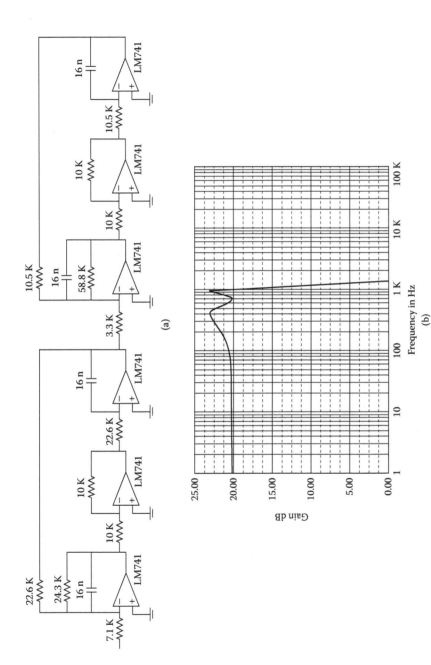

FIGURE 4.14 (a) LP Chebyshev 3 dB filter, $f_1 = 1$ kHz, $K = 10$; (b) its frequency response.

PROBLEMS

4.1 Design an LP state-variable Butterworth 3 dB at 1 kHz and attenuation 45 dB at 2.5 kHz, with a gain of 10.

4.2 Design an LP state-variable Chebyshev 2 dB filter at 350 Hz and attenuation 35 dB at 700 Hz, and with a gain of 1.

4.3 Design a BP state-variable filter with a center frequency 100 Hz, $Q = 20$, $K = 10$, and $n = 4$.

4.4 Figure P4.1 shows the state-variable filter with four op-amps. With this circuit the gain and Q are independent of each other.

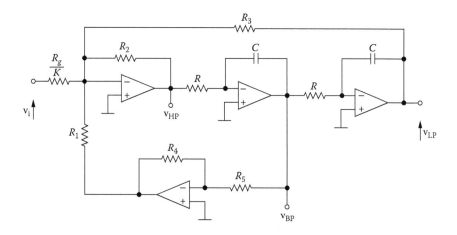

For $R_1 = R_g = R_2 = R_5 = R = 1$ W, $C = 1$ F, and $w_c = w_1 = w_2 = w_0 = 1$ rad/s:
(a) Find the transfer function of the LPF.
(b) Find the formulas to design the filter.

4.5 For Figure P4.1:
(a) Find the transfer function of the HPF.
(b) Find the formulas to design the filter.

4.6 For Figure P4.1:
(a) Find the transfer function of the BPF.
(b) Find the formulas to design the filter.

4.7 Design an LP biquad Butterworth filter at 1 kHz at 3 dB and attenuation 40 dB at 3.5 kHz, with a gain of 10.

4.8 Design a BP biquad filter with $f_0 = 100$ Hz, $Q = 40$, and $K = 10$.

4.9 Design a Butterworth LP state-variable filter with four op-amps (Problem 4.4) at 100 Hz at 3 dB and attenuation 40 dB at 350 Hz and a gain of 10.

4.10 Design a Chebyshev HP state-variable filter with four op-amps at 1 kHz and 30 dB at 500 Hz, with a gain of 1.

4.11 Design a BP state-variable filter with four op-amps at 1 kHz, $Q = 35$, and a gain of 5.

4.12 Prove that the following filter can be used to realize the general biquadratic function (Figure P4.2):

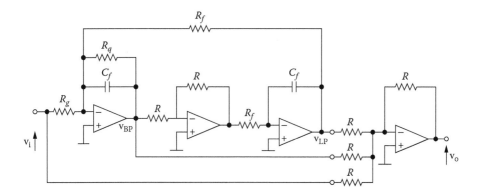

$$H(s) = -K \frac{s^2 + b_1 s + b_0}{s^2 + a_1 s + a_0}$$

5 Sensitivity

5.1 INTRODUCTION

A desired function is realized by interconnecting electrical components of carefully chosen values in a filter network. These components are subject to change due to variations in *temperature, humidity, aging*, and *tolerances* in manufacturing. In order to measure the change in filter performance due to drift or change in component values, we use the *sensitivity* concept.

The basis for all modern-day sensitivity analysis methods is rooted in the work of W.H. Bode. S.J. Mason further expanded Bode's definition, and it is Mason's definition that is most often cited, and hence used here.

Sensitivity is a measure of deviation in some performance characteristic of the circuit due to some change in the nominal value of one or more of the elements of the filter. Low-sensitivity circuits are naturally preferred over high-sensitivity circuits. The *relative sensitivity* is defined as

$$S_x^H = \frac{x}{H}\frac{\partial H}{\partial x} = \frac{\partial \ln H}{\partial \ln x} \tag{5.1}$$

where H is the system transfer function and x is the parameter or network element that is causing H to change.

In general, the network function will be a ratio of polynomials in s, such that

$$H(s,x) = \frac{N(s,x)}{D(s,x)} \tag{5.2}$$

and

$$\frac{dH}{dx} = \frac{D(s,x)N'(s,x) - N(s,x)D'(s,x)}{[D(s,x)]^2} \tag{5.3}$$

where

$$N'(s,x) = \frac{\partial N(s,x)}{\partial x}$$

$$D'(s,x) = \frac{\partial D(s,x)}{\partial x} \tag{5.4}$$

169

Hence,

$$S_x^H = \frac{xD(s,x)}{N(s,x)} \cdot \frac{D(s,x)N'(s,x) - N(s,x)D'(s,x)}{[D(s,x)]^2} \qquad \therefore$$

$$S_x^H = \frac{x}{N(s,x)} \cdot \left[\frac{D(s,x)N'(s,x) - N(s,x)D'(s,x)}{D(s,x)} \right] \qquad (5.5)$$

$$S_x^H = x \left[\frac{N'(s,x)}{N(s,x)} - \frac{D'(s,x)}{D(s,x)} \right] \qquad (5.6)$$

EXAMPLE 5.1

Determine S_x^H for the network function:

$$H(s,x) = \frac{K}{s^2 + xs + 9}$$

where K is some constant, and x will be assumed to have a nominal value of unity. Then,

$$S_x^H = x \left[\frac{0}{K} - \frac{s}{s^2 + xs + 9} \right] \qquad \therefore$$

$$S_x^H = -\frac{xs}{s^2 + xs + 9}$$

At a frequency $\omega = 1$ rad/s, we have:

$$S_x^H = \frac{-j}{-1 + j + 9} = \frac{-j}{8 + j} \qquad \therefore$$

$$S_x^H = \frac{-j(8 - j)}{8^2 + 1} = \frac{-1 - j8}{65} = 0.015 - j0.123$$

The interpretation of the answer is that the real part of the sensitivity specifies a normalized change in the magnitude of the given network function, whereas the imaginary part indicates a change in the argument (phase) of the given network function.

Consequently, the normalized magnitude change of S_x^H with respect to normalized change in x is 0.015, and the phase change with respect to a normalized change in x is 0.123 radians.

5.2 SOME GENERAL PROPERTIES

We next formulate some general properties of the sensitivity function from its definition (Equation 5.1). One form that we will encounter most often is

$$H = h_1^a h_2^b h_3^c \tag{5.7}$$

The natural logarithm of this function is

$$\ln H = a \ln h_1 + b \ln h_2 + c \ln h_3 \tag{5.8}$$

If we differentiate this expression with respect to $\ln h_1$, we have:

$$S_{h_1}^H = \frac{\partial \ln H}{\partial \ln h_1} = a \tag{5.9}$$

Similarly, we have:

$$S_{h_2}^H = b \tag{5.10}$$

$$S_{h_3}^H = c \tag{5.11}$$

If H is expressed as the product of two functions H_1 and H_2, then

$$H = H_1 H_2 \tag{5.12}$$

$$S_x^H = \frac{\partial \ln H}{\partial \ln x} = \frac{\partial \ln(H_1 \cdot H_2)}{\partial \ln x} = \frac{\partial \ln H_1}{\partial \ln x} + \frac{\partial \ln H_2}{\partial \ln x} \tag{5.13}$$

showing that

$$S_x^{H_1 H_2} = S_x^{H_1} + S_x^{H_2} \tag{5.14}$$

Likewise, we can show that

$$S_x^{H_1/H_2} = S_x^{H_1} - S_x^{H_2} \tag{5.15}$$

Other useful relationships are the following:

$$S_x^H = -S_x^{1/H} \tag{5.16}$$

$$S_x^{cH} = S_x^H \quad (c \text{ is independent of } x) \tag{5.17}$$

$$S_x^{H_1+H_2} = \frac{H_1 S_x^{H_1} + H_2 S_x^{H_2}}{H_1 + H_2} \tag{5.18}$$

$$S_x^{H^n} = n S_x^H \tag{5.19}$$

$$S_{x^n}^H = \frac{1}{n} S_x^H \tag{5.20}$$

EXAMPLE 5.2

Perform sensitivity analysis for the low-pass Sallen–Key filter of Figure 2.9.

$$H(s) = \frac{K \dfrac{G_1 G_2}{C_1 C_2}}{s^2 + \left[\dfrac{G_1 + G_2}{C_2} + \dfrac{(1-K)G_2}{C_1}\right]s + \dfrac{G_1 G_2}{C_1 C_2}} \tag{5.21}$$

$$\omega_1^2 = b = \frac{G_1 G_2}{C_1 C_2} = R_1^{-1/2} R_2^{-1/2} C_1^{-1/2} C_2^{-1/2} \tag{5.22}$$

$$B = a\omega_1 = \frac{\omega_1}{Q} = \frac{G_1 + G_2}{C_2} + \frac{(1-K)G_2}{C_1} \qquad \therefore$$

$$B = \frac{\omega_1}{Q} = \frac{1}{R_1 C_2} + \frac{1}{R_2 C_2} + \frac{1-K}{R_2 C_1} \tag{5.23}$$

$$k = K R_1^{-1} R_2^{-1} C_1^{-1} C_2^{-1} \tag{5.24}$$

where $K = 1 + R_b/R_a$

For convenience, we write

$$S_{x_1,x_2,\ ,x_q}^H = S_{x_1}^H = S_{x_2}^H = \ = S_{x_q}^H \tag{5.25}$$

Then, from Equation (5.7) we obtain

$$S_{R_1,R_2,C_1,C_2}^k = -1 \tag{5.26}$$

$$S_K^k = 1 \tag{5.27}$$

$$S_{R,R_2,C_1,C_2}^{\omega_1} = -\frac{1}{2} \tag{5.28}$$

$$S_K^{\omega_1} = S_{R_a}^{\omega_1} = S_{R_b}^{\omega_1} = 0 \tag{5.29}$$

To compute the Q sensitivity, we turn to Equation (5.15), which states that

$$S_x^Q = S_x^{\omega_1} - S_x^B \tag{5.30}$$

where

$$S_x^B = \frac{x}{B} \cdot \frac{\partial B}{\partial x} = x \frac{Q}{\omega_1} \cdot \frac{\partial B}{\partial x} \tag{5.31}$$

We begin by computing Q sensitivity with respect to R_1. For this we differentiate Equation (5.23) with respect to R_1 to give

$$\frac{\partial B}{\partial R_1} = -\frac{1}{R_1^2 C_1}$$

Substituting this in Equation (5.31) in conjunction with Equation (5.23) yields

$$S_{R_1}^B = R_1 Q (R_1 R_2 C_1 C_2)^{-1/2} \left(-\frac{1}{R_1^2 C_1} \right) = -Q R_1^{-1/2} R_2^{-1/2} C_1^{-1/2} C_2^{-1/2} \tag{5.32}$$

giving from Equation (5.15)

$$S_{R_1}^Q = -\frac{1}{2} + Q \sqrt{\frac{R_2 C_2}{R_1 C_1}} \tag{5.33}$$

Similarly, we obtain

$$S_{R_2}^Q = -\frac{1}{2} + Q \left[\sqrt{\frac{R_1 C_2}{R_2 C_1}} + (1-K) \sqrt{\frac{R_1 C_1}{R_2 C_2}} \right] \tag{5.34}$$

$$S_{C_1}^Q = -\frac{1}{2} + Q \left[\sqrt{\frac{R_1 C_2}{R_2 C_1}} + \sqrt{\frac{R_2 C_2}{R_1 C_1}} \right] \tag{5.35}$$

$$S_{C_2}^Q = -\frac{1}{2} + (1-K) Q \sqrt{\frac{R_1 C_1}{R_2 C_2}} \tag{5.36}$$

$$S_K^Q = KQ \sqrt{\frac{R_1 C_1}{R_2 C_2}} \tag{5.37}$$

$$S_{R_a}^Q = -S_{R_b}^Q = (1-K)Q\sqrt{\frac{R_1 C_1}{R_2 C_2}} \tag{5.38}$$

$$S_{R_a}^k = -S_{R_b}^k = \frac{1-K}{K} \tag{5.39}$$

All of these sensitivity functions have been derived for general values of the filter parameters. Once a specific circuit design is chosen, these parameters are known, and the numerical values sensitivities can be ascertained. As a specific example, consider the normalized Butterworth transfer function of

$$H(s) = \frac{2}{s^2 + \sqrt{2}\,s + 1} \tag{5.40}$$

giving from Equation (5.21)

$$K = 2 \qquad \omega_1 = 1 \text{ rad/s} \qquad Q = 0.707 \tag{5.41}$$

This response can be realized by a Sallen–Key low-pass filter with

$$R_1 = R_2 = 1 \ \Omega, \quad C_1 = C_2 = 1 \ \text{F}, \quad R_a = 10 \ \Omega, \quad R_b = 5.86 \ \Omega \tag{5.42}$$

equal component configuration.
Using these parameter values, we obtain

$$S_{R_1}^Q = -S_{R_2}^Q = 0.207$$

$$S_{C_1}^Q = -S_{C_2}^Q = 0.914$$

$$S_K^Q = 1.121$$

$$S_{R_a}^Q = -S_{R_b}^Q = -0.414$$

$$S_{R_a}^k = -S_{R_b}^k = -0.369$$

5.3 MAGNITUDE AND PHASE SENSITIVITIES

To compute the sensitivity functions for the magnitude and phase functions, we express the transfer function in polar form and substitute s by jw to give

$$H(j\omega) = |H(j\omega)|e^{j\phi(\omega)} \tag{5.43}$$

Hence, the sensitivity function becomes [Equation (5.1)]

$$S_x^{H(j\omega)} = \frac{x}{H(j\omega)} \cdot \frac{\partial}{\partial x}\left[|H(j\omega)|e^{j\phi(\omega)}\right] \tag{5.44}$$

which can be expanded by making use of the product rule for differentiation of a product to give

$$S_x^{H(j\omega)} = \frac{x}{|H(j\omega)|} \cdot \frac{\partial |H(j\omega)|}{\partial x} + jx\frac{\partial \phi(\omega)}{\partial x}$$

$$S_x^{H(j\omega)} = S_x^{|H(j\omega)|} + j\phi(\omega)S_x^{\phi(\omega)} \tag{5.45}$$

or

$$S_x^{|H(j\omega)|} = \operatorname{Re} S_x^{H(j\omega)} \tag{5.46}$$

$$S_x^{\phi(\omega)} = \frac{1}{\phi(\omega)} \operatorname{Im} S_x^{H(j\omega)} \tag{5.47}$$

These equations state that the magnitude and phase sensitivity of a transfer function with respect to an element are simply related to the real and imaginary parts of the transfer function sensitivity with respect to the same element.

EXAMPLE 5.3

Find the sensitivity of the low-pass transfer function

$$H(s) = \frac{K\omega_1^2}{s^2 + \dfrac{\omega_1}{Q}s + \omega_1^2} \tag{5.48}$$

The sensitivity function is found from Equation (.5.1) to be

$$S_Q^{H(s)} = \frac{Q}{H(s)} \cdot \frac{\partial H(s)}{\partial Q} = \frac{\dfrac{\omega_1}{Q}s}{s^2 + \dfrac{\omega_1}{Q}s + \omega_1^2} \tag{5.49}$$

To compute the sensitivity function of $|H(j\omega)|$ with respect to Q, we apply Equation (5.46) by first substituting s by jw in Equation (5.49) and then taking the real part. The result is given by

$$S_Q^{|H(j\omega)|} = \operatorname{Re} S_Q^{H(j\omega)} = \frac{(\omega/\omega_1)^2}{Q^2[1-(\omega/\omega_1)^2]^2 + (\omega/\omega_1)^2} \tag{5.50}$$

At $\omega = \omega_1$, the preceding equation becomes

$$S_Q^{|H(j\omega)|}(j\omega_1) = 1 \tag{5.51}$$

The above equation is the sensitivity function of $|H(j\omega)|$ due to the variations in the cutoff frequency ω_1. Applying Equation (5.1) yields

$$S_{\omega_1}^{H(s)} = \frac{2s^2 + \left(\dfrac{\omega_1}{Q}\right)s}{s^2 + \left(\dfrac{\omega_1}{Q}\right)s + \omega_1^2} \tag{5.52}$$

giving

$$S_{\omega_1}^{|H(j\omega)|} = \text{Re } S_{\omega_1}^{H(j\omega)} = \frac{\left(\dfrac{\omega}{\omega_1}\right)^2 \left[2\left(\dfrac{\omega}{\omega_1}\right)^2 + \dfrac{1}{Q^2} - 2\right]}{\left[1 - \left(\dfrac{\omega}{\omega_1}\right)^2\right]^2 + \left(\dfrac{\omega}{Q\omega_1}\right)^2} \tag{5.53}$$

At $\omega = \omega_1$, the preceding equation becomes

$$S_{\omega_1}^{|H(j\omega)|}(j\omega_1) = 1 \tag{5.54}$$

5.4 ROOT SENSITIVITY

The location of the poles of an active filter determines the stability of the network. Therefore, it is important to know the manner in which these poles vary as some of the network elements change. As poles of a network function are roots of a polynomial, it suffices to examine how the roots of a polynomial change as the value of an element changes.

The **root sensitivity** is defined as the ratio of the change in a root to the fractional change in an element for the situation when all changes concerned are differentially small.

Thus, if s_j is a root of a polynomial $D(s)$, the root sensitivity of s_j with respect to an element x is defined by the equation

$$S_x^{-s_j} = \frac{\partial s_j}{\partial x / x} = x \frac{\partial s_j}{\partial x} \tag{5.55}$$

If the transfer function has only poles, i.e.,

$$H(s) = \frac{K}{\displaystyle\prod_{j=1}^{k}(s - p_j)^{k_j}} \tag{5.56}$$

In term of the logarithms, the preceding equation becomes

$$\ln H(s) = \ln K - \sum_{j=1}^{k} k_j \ln (s - p_j) \tag{5.57}$$

Taking partial derivatives with respect to x on both sides yields

$$\frac{\partial \ln H(s)}{\partial x} = \frac{1}{K} \cdot \frac{\partial K}{\partial x} + \sum_{j=1}^{k} \frac{k_j \dfrac{\partial p_j}{\partial x}}{s - p_j} \tag{5.58}$$

which can be put in the form

$$\frac{x}{H(s)} \cdot \frac{\partial H(s)}{\partial x} = \frac{x}{K} \cdot \frac{\partial K}{\partial x} + \sum_{j=1}^{k} \frac{k_j x \dfrac{\partial p_j}{\partial x}}{s - p_j} \tag{5.59}$$

This gives a relationship between the function sensitivity and pole sensitivity:

$$S_x^H = S_x^K + \sum_{j=1}^{k} \frac{k_j S_x^{p_j}}{s - p_j} \tag{5.60}$$

EXAMPLE 5.4

Consider the transfer function of the second-order active filter

$$H(s) = \frac{K}{s^2 + \dfrac{\omega_1}{Q} s + \omega_1^2} \tag{5.61}$$

In the only case of real interest, the poles are complex ($Q > 0.5$), so that p_1 and p_2 are conjugate, i.e., $p_2 = p_1^*$, with

$$p_1 = -\omega_1 \left(\frac{1}{2Q} - j\sqrt{1 - \frac{1}{4Q^2}} \right) \tag{5.62}$$

From the preceding equation and Equation (5.55), assuming that Q and ω_1 are functions of x, we calculate

$$S_x^{p_1} = p_1 \left(S_x^{\omega_1} - j \frac{S_x^Q}{\sqrt{4Q^2 - 1}} \right) \tag{5.63}$$

and $Q_x^{p_2} = (Q_x^{p_1})^*$. From the preceding equation we observe that the location of a pole is $\sqrt{4Q^2 - 1} \cong 2Q$ times more sensitive to variations in ω_1 than it is to variation in Q.

Having established an expression for the pole sensitivity $S_x^{p_1}$ in Equation (5.63), we now investigate its effect on the passband of the biquadratic transfer function [Equation (5.61)]. To this end we evaluate Equation (5.58) for $H(s)$ under the assumption that the parameter x does not affect the gain constant:

$$S_x^{H(s)} = \frac{S_x^{p_1}}{s - p_1} + \frac{(S_x^{p_1})^*}{s - p_1^*} \tag{5.64}$$

If we express the right-hand side in terms of its common denominator $s^2 + (\omega_1 / Q)s + \omega_1^2$ and use Equation (5.57), we obtain

$$S_x^{H(s)} = -\frac{\left(2\omega_1^2 + \dfrac{\omega_1}{Q} s\right) S_x^{\omega_1} - \dfrac{\omega_1}{Q} s S_x^{Q}}{s^2 + \dfrac{\omega_1}{Q} s + \omega_1^2} \tag{5.65}$$

Finally, if we use Equation (5.46), we can derive the magnitude sensitivity

$$S_x^{|H(j\omega)|} = -\frac{2(1-\omega_n^2) + \left(\dfrac{\omega_n}{Q}\right)^2}{(1-\omega_n^2)^2 + \left(\dfrac{\omega_n}{Q}\right)^2} S_x^{\omega_1} + \frac{\left(\dfrac{\omega_n}{Q}\right)^2}{(1-\omega_n^2)^2 + \left(\dfrac{\omega_n}{Q}\right)^2} S_x^{Q}$$

or

$$S_x^{|H(j\omega)|} = S_{\omega_1}^{|H|} S_x^{\omega_1} + S_Q^{|H|} S_x^{Q} \tag{5.66}$$

where $\omega_n = \omega / \omega_1$ is the normalized frequency, where

$$S_{\omega_1}^{|H|} = -\frac{2(1-\omega_n^2) + \left(\dfrac{\omega_n}{Q}\right)^2}{(1-\omega_n^2)^2 + \left(\dfrac{\omega_n}{Q}\right)^2} \tag{5.67}$$

$$S_Q^{|H|} = \frac{\left(\dfrac{\omega_n}{Q}\right)^2}{(1-\omega_n^2)^2 + \left(\dfrac{\omega_n}{Q}\right)^2} \tag{5.68}$$

A better appreciation of the meaning of Equations (5.67) and (5.68) is obtained from the plot of $S_{\omega_1}^{|H|}$ and $S_Q^{|H|}$. As shown in Figure 5.1, both $S_{\omega_1}^{|H|}$ and $S_Q^{|H|}$ are strong functions of frequency.

From the aforementioned equations:

At $\omega = \omega_1$, we have

$$\max\left(S_Q^{|H|}\right) = 1 \tag{5.69}$$

and for large Q

$$\max\left(S_{\omega_1}^{|H|}\right) \cong \frac{Q}{1+\dfrac{1}{Q}} \qquad \text{at} \qquad \omega \cong \omega_1\left(1+\frac{1}{2Q}\right) \tag{5.70}$$

$$\min\left(S_{\omega_1}^{|H|}\right) \cong -\frac{Q}{1+\dfrac{1}{Q}} \qquad \text{at} \qquad \omega \cong \omega_1\left(1+\frac{1}{2Q}\right) \tag{5.71}$$

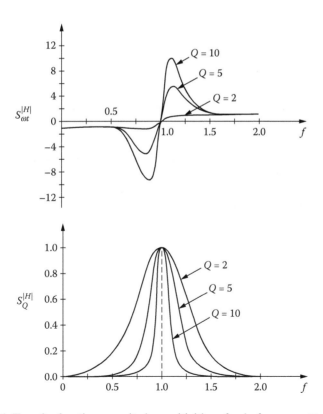

FIGURE 5.1 Transfer function magnitude sensitivities of pole frequency w_1 and quality factor Q.

Note that the extreme values of $S_{\omega_1}^{|H|}$ occur approximately at the 3 dB frequencies of high-Q second-order functions.

For good practical second-order sections with high values of Q, it is more important to pay attention to low values of $S_x^{\omega_1}$ than to small values of S_x^Q.

PROBLEMS

5.1. For the circuit shown in Figure P5.1 (assume an ideal op-amp):
    (a) Find the transfer function.
    (b) To synthesize a normalized Butterworth two-pole transfer function, namely,

$$H(s) = \frac{10}{s^2 + \sqrt{2}\,s + 1}$$

    what are the design equations?
    (c) Find the sensitivity with respect to the gain at $s = j$.

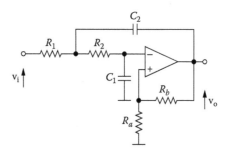

5.2 For the circuit shown in Figure P5.2, assume an ideal op-amp:
    (a) Find the transfer function.
    (b) Determine the various sensitivity functions $S_x^{\omega_0}$ and S_x^Q, where x denotes the R's and the C's, respectively.

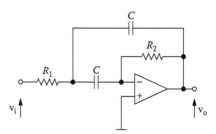

5.3 For the circuit shown in Figure P5.3 (assume an ideal op-amp):
    (a) Find the transfer function.
    (b) Determine the various sensitivity functions S_x^H, where x denotes the R's and the C's, respectively.

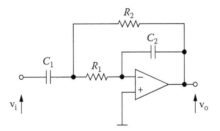

5.4 For the circuit shown in Figure P5.4:
(a) Find the transfer function.
(b) Calculate all S_x^H, where the x denotes the RLC components.

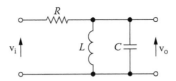

5.5 For the circuit shown in Figure P5.5, assume an ideal op-amp:
(a) Find the transfer function.
(b) Determine the sensitivities S_x^a, $S_x^{\omega_1}$, S_x^K, where x denotes the R's and the C's, respectively.

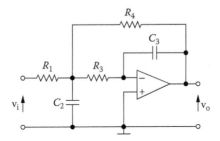

5.6 For the circuit shown in Figure P5.6 (assume an ideal op-amp):
(a) Find the transfer function.
(b) Determine the sensitivities $S_x^{\omega_2}$, S_x^a, $S_{C_1}^K$, $S_{C_4}^K$, where x denotes the R's and the C's, respectively.

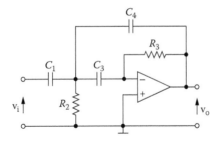

5.7 For the circuit shown in Figure P5.7 (assume an ideal op-amp):
     (a) Find the transfer function.
     (b) Determine the sensitivities $S_x^{\omega_0}$, S_x^Q, where x denotes the R's and the
     C's, respectively.

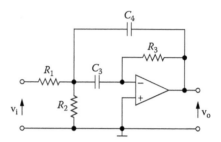

5.8 For the circuit shown in Figure P5.8 (assume an ideal op-amp):
     (a) Find the transfer function.
     (b) Determine the sensitivities $S_x^{\omega_2}$, S_x^a, where x denotes the R's and the
     C's, respectively.

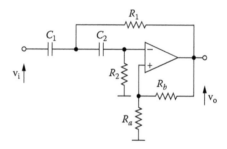

5.9 For the circuit shown in Figure P5.9 (assume an ideal op-amp):
     (a) Find the transfer function.
     (b) Determine the sensitivities $S_x^{\omega_0}$, S_x^Q, where x denotes the R's and the
     C's, respectively.

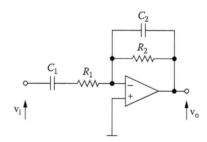

6 Filters with GIC

6.1 INTRODUCTION

Impedance converters are active RC circuits designed to simulate frequency-dependent elements such as *impedances* for use in active filter synthesis. Among the various configurations is the *generalized impedance converter* (GIC), which can be used to simulate inductances.

This topology allows one to easily realize active filters beginning from a passive filter design. In addition, the GIC filter provides extremely low distortion and noise, at a reasonable cost. Compared with more familiar feedback techniques, such as Sallen–Key filter topologies, the GIC filter has superior noise gain characteristics, making it particularly suitable for *audio* and *DSP type applications*.

6.2 GENERALIZED IMPEDANCE CONVERTERS

GICs are electronic circuits used to convert one impedance into another impedance. GICs provide a way to get the advantages of passive circuits (the transfer function is relative insensitive to variations in the values of the resistances and inductances) without the disadvantages of inductors (which are frequently large, heavy, expensive, and nonlinear). The GIC (Figure 6.1) converts the impedance $Z_2(s)$ to the impedance $Z_1(s)$.

$$Z_1(s) = K(s)Z_2(s) \qquad (6.1)$$

Figure 6.2 shows the way to implement a GIC using op-amps. The equivalent Z of this circuit is V/I. Because each op-amp has $v^+ = v^-$, the voltage at the input nodes of both op-amps is v. By Ohm's law, we have:

$$I = (V - V_{02})Y_{11} \qquad (6.2)$$

node v_1

$$-Y_{12}V_{02} + (Y_{12} + Y_{13})V - Y_{13}V_{01} = 0 \qquad (6.3)$$

node v_2

$$-Y_{14}V_{01} + (Y_{14} + Y_{15})V = 0 \qquad (6.4)$$

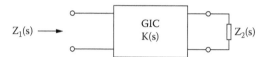

FIGURE 6.1 The GIC converter.

Eliminating V_{01} and V_{02}, and solving for the ratio V/I, we have:

$$V_{01} = \frac{Y_{14} + Y_{15}}{Y_{14}} V \tag{6.5}$$

From Equations (6.5) and (6.3), we have:

$$Y_{12} V_{02} = (Y_{12} + Y_{13}) V - \frac{Y_{13}(Y_{14} + Y_{15})}{Y_{14}} V \qquad \therefore$$

$$V_{02} = \frac{Y_{12} Y_{14} - Y_{13} Y_{15}}{Y_{12} Y_{14}} V \tag{6.6}$$

From Equations (6.2) and (6.6), we get

$$I = \left[V - \frac{Y_{12} Y_{14} - Y_{13} Y_{15}}{Y_{12} Y_{14}} V \right] Y_{11} \qquad \therefore$$

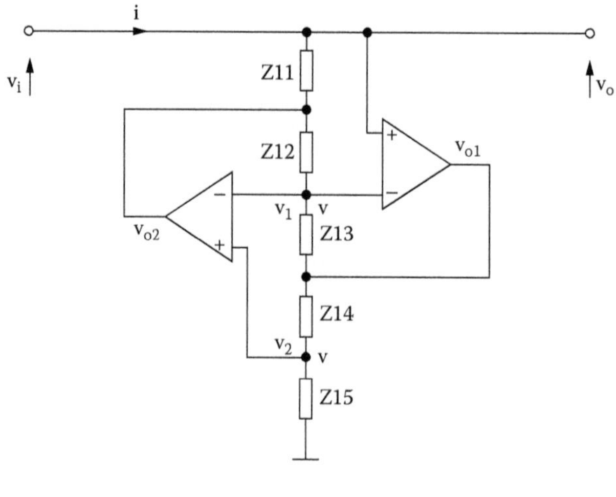

FIGURE 6.2 Generalized impedance converter (GIC).

$$Z = \frac{V}{I} = \frac{Y_{12}Y_{14}}{Y_{11}Y_{13}Y_{15}} \qquad \therefore$$

$$Z = \frac{Z_{11}Z_{13}Z_{15}}{Z_{12}Z_{14}} \tag{6.7}$$

1. If Z_{11}, Z_{13}, Z_{14}, and Z_{15} are resistances and $Z_{12} = 1/sC$, Equation (6.7) gives

$$Z = \frac{R_{11}R_{13}R_{15}}{R_{14}} sC = sL \tag{6.8a}$$

where

$$L = \frac{R_{11}R_{13}R_{15}}{R_{14}} C \tag{6.8b}$$

indicating that the circuit simulates a grounded inductance (Figure 6.3). If desired, this inductance can be adjusted by varying one of the resistances.

2. If Z_{12}, Z_{13}, and Z_{14} are resistances and Z_{15} are capacitances, Equation (6.7) gives

$$Z = \frac{R_{13}}{s^2 R_{13}R_{14}C_{11}C_{15}} = -\frac{1}{\omega^2 D} \tag{6.9a}$$

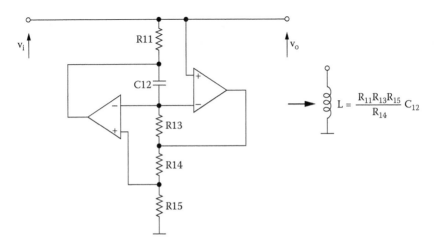

FIGURE 6.3 Inductance simulator.

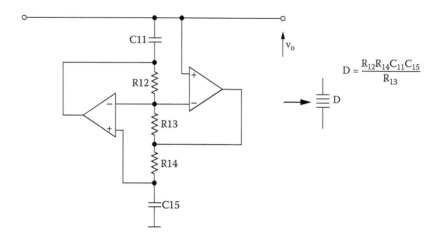

FIGURE 6.4 Realization of D element.

where

$$D = \frac{R_{12}R_{14}C_{11}C_{15}}{R_{13}} \qquad (6.9b)$$

The circuit now simulates a *grounded frequency-dependent negative resistance* (FDNR or D element) (Figure 6.4). The D element can be adjusted by varying one of the resistances.

If all the impedances of an LRC filter are multiplied by $1/s$, the transfer function remains unchanged. This operation is equivalent to impedance-scaling a filter by the factor $1/s$ and should not be confused with the high-pass transformation.

When the elements of a network are impedance-scaled by $1/s$, inductors are transformed into resistors, resistors into capacitors, and capacitors into a D element (Table 6.1). This design technique is very powerful. It enables the designer to design active filters directly from the passive RLC filters, using suitable filter tables or computer programs. The filter is then realized in active form by replacing its inductor with the simulated ones. The resulting filter is expected to have low sensitivity, as its passive counterpart, except for the imperfection in the realization of the inductor using the RC circuit.

TABLE 6.1
The 1/s Impedance Transformation

Element	Impedance	Transformed Element	Impedance
R	R	C	R/s
L	sL	R	L
C	1/sC	D	$1/s^2C$

6.3 LOW-PASS FILTER DESIGN

To design a GIC filter, the starting point is a passive ladder prototype, which is designed using suitable tables or computer programs. The filter is then realized in active form by replacing its inductors with simulated ones. The resulting active network retains the low-sensitivity advantages of its RLC prototype, a feature that makes it suitable for applications with stringent specifications.

EXAMPLE 6.1

A third-order low-pass Butterworth filter must be designed with cutoff frequency 1 kHz.

Solution

From Appendix E we find the normalized RLC filter.

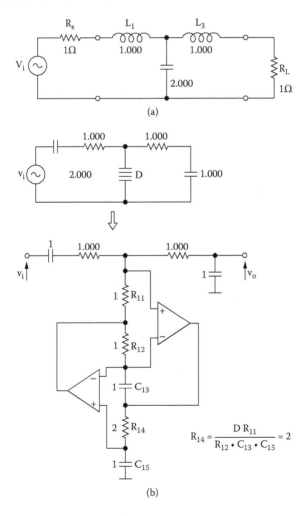

FIGURE 6.5 Third-order LP Butterworth filter using FDNR.

We cannot use GIC simulation because the RLC filter contains floating inductances. This obstacle is overcome by applying the $1/s$ transformation, after which resistances are changed to capacitances, the inductances to resistances, and the capacitances to D elements.

The D elements are realized using the GIC as shown in the following figure:

Frequency and impedance scaling ($FSF = 2\pi \times 10^3$ and for $C = 10$ nF \ $ISF = 15.92 ¥ 10^3$) gives the designed active GIC filter (Figure 6.5a) and its frequency response (Figure 6.5b).

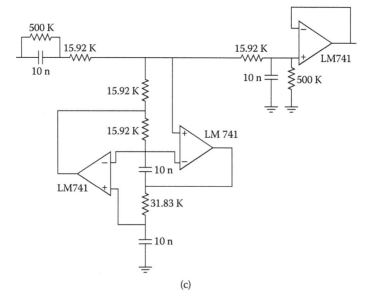

(c)

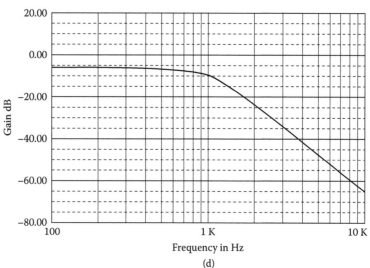

(d)

FIGURE 6.5 (Continued)

We use 500 KW resistance at the input to provide a dc path for the op-amps. To ensure a dc gain of 0.5, this resistance must be counterbalanced by a 500 KW resistance at the output. To avoid loading problems, an output buffer is used. The FDNR can be tuned by adjusting one of its resistances.

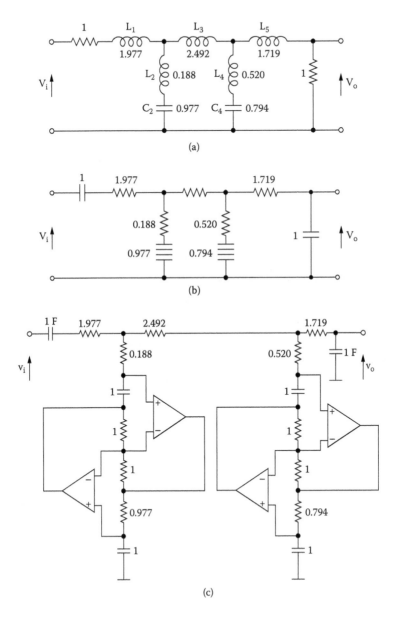

FIGURE 6.6 LP elliptic filter: (a) normalized low-pass RLC filter; (b) circuit after $1/s$ transformation; (c) normalized configuration using GICs for D elements; (d) denormalized filter; (e) frequency response.

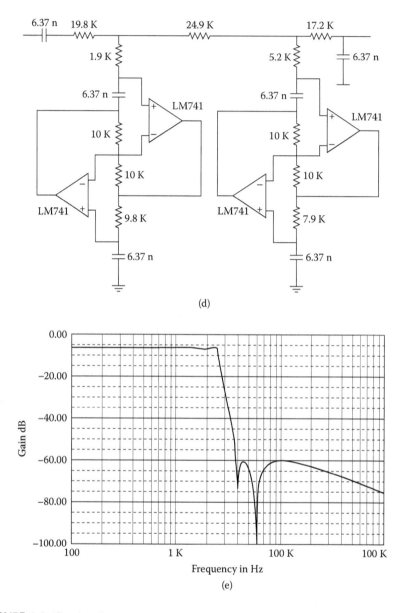

(d)

FIGURE 6.6 (Continued)

EXAMPLE 6.2

A fifth-order low-pass elliptic filter must be designed with $f_1 = 2.5$ kHz, 1 dB pass-band ripple, and $\omega_s = 1.50$.

Solution

From Appendix E we find the normalized RLC filter (Figure 6.6).

$ISF = 10^4$; also, we calculate $FSF = 2\pi f_1 = 2\pi \times 2.5 \times 10^3$

∴

$$k = ISF \times FSF = 5\pi \times 10^7 \quad \therefore \quad C = 6.37 \text{ nF}$$

6.4 HIGH-PASS FILTER DESIGN

An active realization of a ground inductor is particularly suited for the design of active high-pass filters. If a passive RLC low-pass filter is transformed into a high-pass filter, shunt inductors to ground are obtained that can be implemented using GICs. The resulting normalized filter can then be frequency- and impedance-scaled. If R_{15} is made variable, the equivalent inductance can be adjusted.

EXAMPLE 6.3

A fifth-order high-pass Butterworth filter must be designed with cutoff frequency 5 kHz.

Solution

From Appendix E we find the normalized RLC filter, and from this we design the GIC filter (Figure 6.7).
   $ISF = 10^4$; also, we calculate $FSF = 2\pi f_2 = 2\pi \times 10^4$ and $C = C_n/(ISF \times FSF)$.

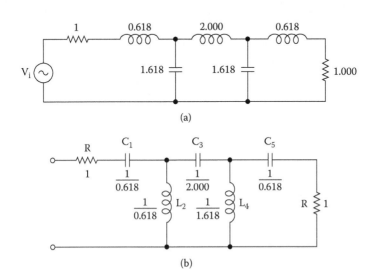

(a)

(b)

FIGURE 6.7 Fifth-order Butterworth HP filter: (a) prototype low-pass filter; (b) transformed high-pass filter; (c) high-pass filter using GIC; (d) frequency- and impedance-scaled filter; (e) its frequency response.

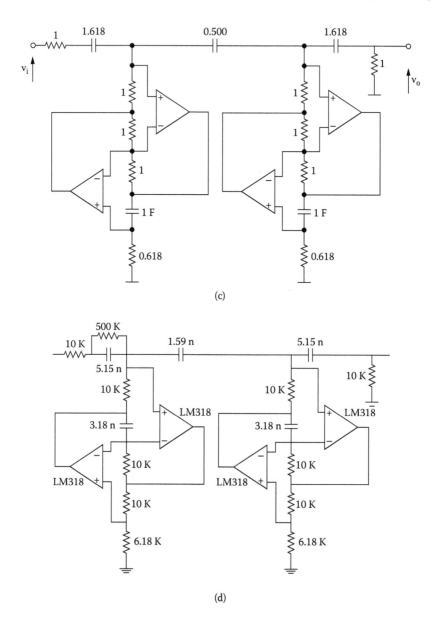

FIGURE 6.7 (Continued)

EXAMPLE 6.4

A third-order high-pass elliptic filter 0.1 dB and $\omega_s = 1.5$ must be designed with cutoff frequency 10 kHz.

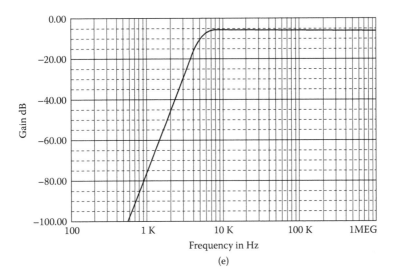

(e)

FIGURE 6.7 (Continued)

Solution

From Appendix E we find the prototype RLC filter from which we design the GIC filter (Figure 6.8).

$$ISF = 10^4, \quad FSF = 2\pi f_2 = 2\pi \times 3 \times 10^4 = 6\pi \times 10^4 \quad \therefore \quad ISF \times FSF = 6\pi \times 10^8$$

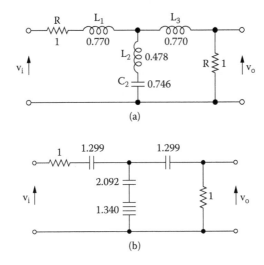

FIGURE 6.8 Elliptic high-pass filter: (a) RLC normalized low-pass filter; (b) transformed high-pass filter; (c) high-pass filter using GIC; (d) frequency- and impedance-scaled filter; (e) frequency response.

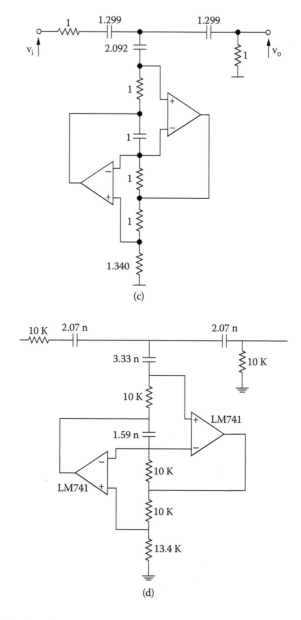

FIGURE 6.8 (Continued)

6.5 NARROW-BAND BAND-PASS FILTER DESIGN

Figure 6.9 shows the narrow-band band-pass RLC filter. From this filter, we have:
node v_o

$$-GV_i + \left(sC + \frac{1}{sL} + G \right) V_o = 0 \qquad \therefore$$

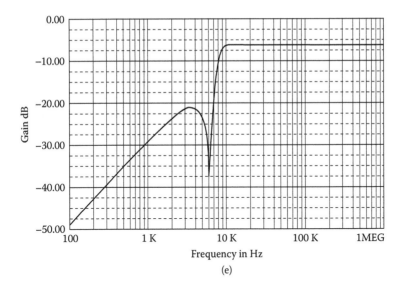

Frequency in Hz

(e)

FIGURE 6.8 (Continued)

$$H(s) = \frac{V_o}{V_i} = \frac{G}{sC + \dfrac{1}{sL} + G} = \frac{sGL}{s^2 LC + sLG + 1} \qquad \therefore$$

$$H(s) = \frac{\dfrac{1}{Q}\left(\dfrac{s}{\omega_o}\right)}{\left(\dfrac{s}{\omega_o}\right)^2 + \dfrac{1}{Q}\left(\dfrac{s}{\omega_o}\right) + 1} \tag{6.10}$$

where

$$\omega_o^2 = \frac{1}{LC} \tag{6.11}$$

and

$$Q = \omega_o RC \tag{6.12}$$

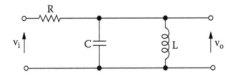

FIGURE 6.9 RLC narrow-band band-pass filter.

EXAMPLE 6.5

Design a narrow-band band-pass filter with GIC circuit with $f_o = 1 \text{ kHz}$ and $Q = 10$.

Solution

(a) Design the RLC normalized filter.
For $\omega_o = 1$ rad/s, choose C = 1 F ∴ L = 1 H and Q = 10, R = 10 W.
(b) We specify the components for the GIC circuit.
We accept equal resistors and capacitors:

$$R_{11} = R_{13} = R_{14} = 1 \Omega \text{ and } C = C_{12} = 1 F$$

$$ISF = 10^4, \quad FSF = \frac{\omega_o}{\omega_n} = 2\pi f_o = 2\pi \times 10^3 \quad \therefore \quad k = ISF \times FSF = 2\pi \times 10^7$$

∴

$$R = ISF \times R_n = 10^4 \times 10 \, \Omega = 100 \text{ k}\Omega$$

$$R_{11} = R_{13} = R_{14} = R_{15} = ISF \times R_{11n} = 10^4 \times 1 \, \Omega = 10 \text{ k}\Omega$$

$$C = C_{12} = \frac{C_n}{k} = \frac{1}{2\pi \times 10^7} = 15.9 \text{ nF}$$

Figure 6.10 depicts the designed narrow-band band-pass filter with GIC.

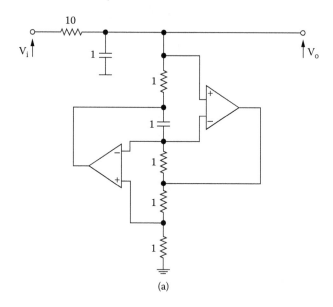

(a)

FIGURE 6.10 (b) Narrow-band band-pass active filter with GIC circuit; (c) its frequency response.

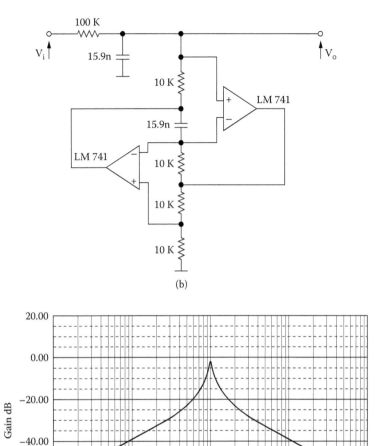

(b)

(c)

FIGURE 6.10 (Continued)

6.6 NARROW-BAND BAND-REJECT FILTER DESIGN

Figure 6.11 shows the narrow-band band-reject RLC filter. From this filter, we have:

$$H(s) = \frac{V_o}{V_i} = \frac{Z_2}{Z_1 + Z_2} = \frac{sL + \dfrac{1}{sC}}{R + sL + \dfrac{1}{sC}} \qquad \therefore$$

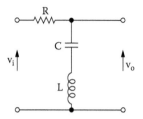

FIGURE 6.11 Narrow-band band-reject RLC filter.

$$H(s) = \frac{\left(\dfrac{s}{\omega_o}\right)^2 + 1}{\left(\dfrac{s}{\omega_o}\right)^2 + \dfrac{1}{Q}\left(\dfrac{s}{\omega_o}\right) + 1} \tag{6.13}$$

where

$$\omega_o^2 = \frac{1}{LC} \tag{6.14}$$

and

$$Q = \frac{\omega_o L}{R} \tag{6.15}$$

EXAMPLE 6.6

Design a narrow-band band-reject filter with GIC circuit with $f_0 = 1\,\text{kHz}$ and $Q = 10$.

Solution

For $\omega_o = 1\,\text{rad/s}$, choose $C = 1\,\text{F}$ and from Equations (6.14) and (6.15), we have $L = 1\,\text{F}$ and $R = 1/Q = 0.1\,\Omega$.

We use equal resistors and capacitors for the GIC circuit:

$$R_{11} = R_{13} = R_{14} = 1\,\Omega, \ C = C_{12} = 1\,\text{F} \ \therefore \ R_{15} = 1\,\Omega$$

$$ISF = 10^4, \ FSF = 2\pi f_o = 2\pi \times 10^3 \ \therefore \ k = ISF \times FSF = 2\pi \times 10^7 \ \therefore$$

$$R_{11} = R_{13} = R_{14} = R_{15} = ISF \times R_{11n} = 10^4 \times 1\,\Omega = 10\,\text{k}\Omega$$

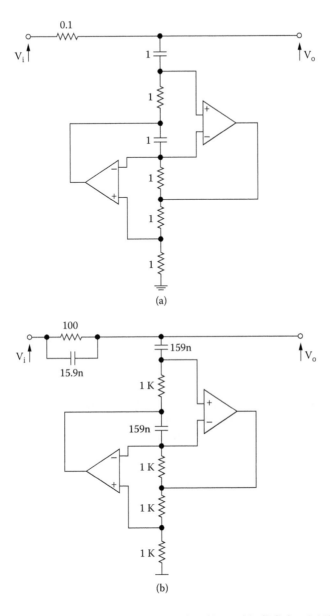

FIGURE 6.12 (b) Narrow-band band-reject active filter with GIC $f_0 = 1$ kHz, $Q = 10$; (c) frequency response.

$$C = C_{12} = \frac{C_n}{k} = \frac{1}{2\pi \times 10^7} = 15.9 \text{ nF}$$

Figure 6.12a depicts the designed filter, and its frequency response is shown in Figure 6.12b.

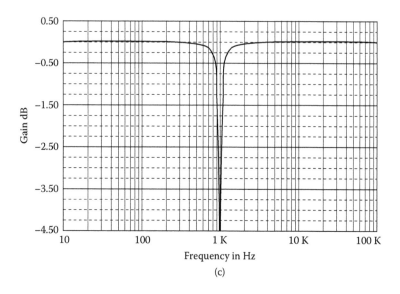

(c)

FIGURE 6.12 (Continued)

PROBLEMS

6.1 Provided $R = \sqrt{L/2C}$, the circuit of Figure P6.1 yields a third-order low-pass Butterworth response with 3 dB frequency $\omega_1 = 1/\sqrt{2LC}$.
(a) Specify suitable components for $w_1 = 1$ rad/s.
(b) Convert the circuit to a GIC realization for $f_1 = 1$ kHz

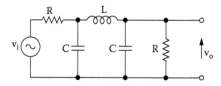

6.2 Provided $R = \sqrt{L/2C}$, the circuit of Figure P6.2 yields a third-order high-pass Butterworth response with 3 dB frequency $\omega_2 = 1/\sqrt{2LC}$.
(a) Specify suitable components for $\omega_2 = 1$ rad/s.
(b) Convert the circuit to a GIC realization for $f_2 = 3$ kHz.

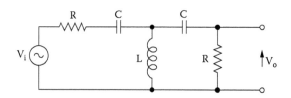

6.3 Using GICs and information of Appendix E, design a seventh-order 1-dB Chebyshev 1-dB ripple passband high-pass filter with $f_2 = 300$ Hz.

6.4 It is desired to design a seventh-order 0.5-dB Chebyshev low-pass filter with $f_1 = 5$ kHz using FDNR implementation.

6.5 Find an FDNR realization for the third-order low-pass Chebyshev 0.5-dB ripple, with $f_1 = 5$ kHz (Figure P6.3).

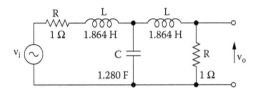

6.6 (a) Find the transfer function of the filter of Figure P6.4:

(b) If the normalized transfer function of the preceding filter is

$$H_n(s) = \frac{1}{s^2 + \sqrt{2}\,s + 1}$$

(c) Find an FDNR realization of this filter for $f_1 = 3$ kHz.

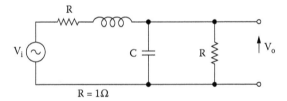

6.7 Find an FDNR realization of the second-order high-pass filter using the results of Problem 6.6 for $f_2 = 300$ Hz.

6.8 Using GICs and information of Appendix E, design a fourth-order Butterworth LP filter with $f_1 = 500$ Hz.

6.9 Using GICs and information of Appendix E, design a seventh-order low-pass elliptic 1-dB passband ripple with $\omega_s = 2.00$ and $f_1 = 10$ kHz.

6.10 Using GICs and information of Appendix E, design a third-order high-pass 0.1-dB passband ripple with $\omega_s = 2.00$ and $f_2 = 1$ kHz.

7 OTA Filters

7.1 INTRODUCTION

An ideal *operational transconductance amplifier* (OTA) is a voltage-controlled current source. What is important and useful about the OTA's transconductance parameter is that it is controlled by an external current, the amplifier current I_b, so that one obtains

$$g_m = \frac{20}{V_i} I_B \tag{7.1}$$

From this externally controlled transconductance, the output current as a function of the applied voltage difference between the two pins labeled v_2 and v_1 is given by

$$i_o = g_m v_d \tag{7.2}$$

or

$$i_o = g_m(v_2 - v_1) \tag{7.3}$$

where g_m is the transconductance parameter provided by the active devices. The transconductance of the OTA is in the range of tens to hundreds of μS (CMOS technology) and up to mS (bipolar technology). The OTA can work in the frequency range of 50 to several MHz. Its circuit model is shown in Figure 7.1a. To avoid loading effects both at the input and at the output, an OTA should have $R_i = R_o = \infty$.

OTAs find applications in their own right. The OTA is a fast device. Moreover, g_m can be varied by changing the bias current by differential transistor pairs, making OTAs suited to electronically programmable functions.

Programmable high-frequency active filters can therefore be achieved by using the OTA. However, single OTA filters may not be suitable for full integration as they contain resistors that require large chip area. In recent years active filters use only OTA and capacitors. These filters are called OTA-C filters. The single OTA filter structures can be converted into integrated OTA-C filters by using OTAs to simulate the resistors.

7.2 SINGLE OTA LP FILTERS WITH THREE PASSIVE COMPONENTS

Figure 7.2a shows an OTA with three impedances, and Figure 7.2b shows its equivalent circuit. From Figure 7.2b we have:

node v_o

$$-Y_2V_1 + (Y_2 + Y_3)V_o - g_m(V_i - V_1) = 0 \qquad (7.4)$$

node v_1

$$-Y_2V_o + (Y_1 + Y_2)V_1 = 0 \qquad (7.5)$$

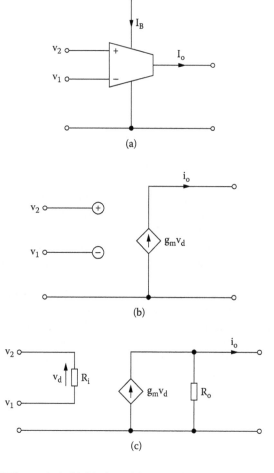

(a)

(b)

(c)

FIGURE 7.1 (a) OTA: symbol, (b) ideal model, (c) equivalent circuit.

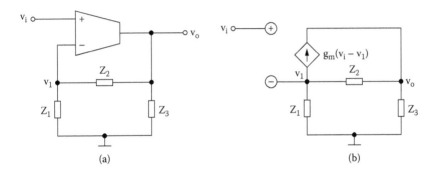

FIGURE 7.2 (a) Single OTA filter with three impedance, (b) equivalent circuit.

From the preceding equations we have:

$$(Y_2 + Y_3)V_o - \frac{(Y_2 - g_m)Y_2}{Y_1 + Y_2}V_o = g_m V_i \qquad \therefore$$

$$H_1(s) = \frac{V_o}{V_i} = \frac{g_m(Y_1 + Y_2)}{Y_1 Y_2 + Y_2 Y_3 + Y_1 Y_3 + g_m Y_2} \tag{7.6}$$

and

$$H_2(s) = \frac{V_1}{V_i} = \frac{g_m Y_2}{Y_1 Y_2 + Y_2 Y_3 + Y_1 Y_3 + g_m Y_2} \tag{7.7}$$

From these expressions we can derive different first- and second-order filters.

7.2.1 FIRST-ORDER LOW-PASS FILTER

Figure 7.3 shows a low-pass filter and its equivalent circuit.

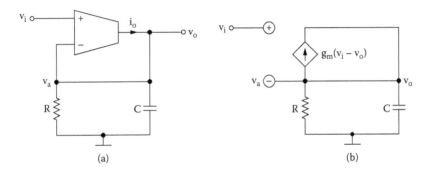

FIGURE 7.3 First-order low-pass filter and its equivalent circuit.

From Figure 7.3b, we have:

$$I_o = g_m(V_i - V_o) \tag{7.8}$$

$$V_o = I_o Z_L \tag{7.9}$$

$$Y_L = sC + G \qquad \therefore$$

$$Z_L = \frac{1}{Y_L} = \frac{1}{sC + G} \tag{7.10}$$

From the preceding equations, we have:

$$V_o = \frac{g_m(V_i - V_o)}{sC + G} \qquad \therefore$$

$$H(s) = \frac{V_o}{V_i} = \frac{g_m}{sC + g_m + G} \qquad \therefore$$

$$H(s) = \frac{K}{1 + \dfrac{s}{\omega_1}} \tag{7.11}$$

where

$$\omega_1 = \frac{1 + g_m R}{RC} \tag{7.12}$$

$$K = \frac{g_m R}{1 + g_m R} \tag{7.13}$$

7.2.2 FIRST-ORDER HIGH-PASS FILTER

Figure 7.4a shows a high-pass first-order filter with three passive components, and Figure 7.4b shows its equivalent circuit.
node v_a

$$(sC + G_1)V_a - sCV_o = 0 \tag{7.14}$$

node v_o

$$-sCV_a + (sC + G_2)V_o - g_m(V_i - V_a) = 0 \tag{7.15}$$

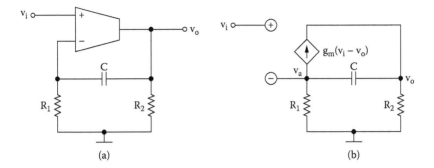

FIGURE 7.4 (a) First-order high-pass OTA filter, (b) its equivalent circuit.

From these equations we have:

$$\frac{(sC+G_1)(sC+G_2)}{sC}V_a+(g_m-sC)V_a=g_mV_i \qquad \therefore$$

$$H(s)=\frac{V_a}{V_i}=\frac{3g_mC}{sC(g_m+G_1+G_2)+G_1G_2} \qquad \therefore$$

$$H(s)=\frac{K\left(\dfrac{s}{\omega_2}\right)}{1+\dfrac{s}{\omega_2}} \tag{7.16}$$

where

$$K=\frac{g_m}{g_m+G_1+G_2}=\frac{g_m}{1+g_mR} \tag{7.17}$$

$$R=R_1//R_2 \tag{7.18}$$

and

$$\omega_2=\frac{G_1G_2}{C(g_m+G_1+G_2)}=\frac{1}{C(g_mR_1R_2+R_1+R_2)} \tag{7.19}$$

7.3 SECOND-ORDER LOW-PASS FILTER

Figure 7.5a shows a second-order low-pass filter, and its equivalent circuit is shown in Figure 7.5b.

From Figure 7.5b we have:
node v_o

$$(sC_1+G)V_o-GV_a=0 \tag{7.20}$$

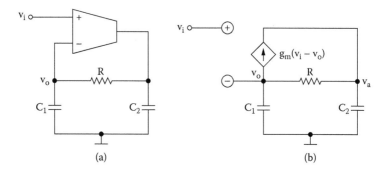

FIGURE 7.5 (a) Second-order low-pass filter; (b) equivalent circuit.

node v_a

$$-GV_o + (sC_2 + G)V_a - g_m(V_i - V_o) = 0 \qquad (7.21)$$

From Equations (7.20) and (7.21) we have:

$$\frac{(sC_1 + G)(sC_2 + G)}{G}V_o + (g_m - G)V_o = g_m V_i \qquad \therefore$$

$$H(s) = \frac{V_o}{V_i} = \frac{g_m G}{s^2 C_1 C_2 + s(C_1 + C_2)G + g_m G} \qquad \therefore$$

$$H(s) = \frac{\dfrac{g_m G}{C_1 C_2}}{s^2 + \dfrac{(C_1 + C_2)G}{C_1 C_2}s + \dfrac{g_m G}{C_1 C_2}} \qquad \therefore$$

$$H(s) = \frac{1}{\left(\dfrac{s}{\omega_1}\right)^2 + \dfrac{(C_1 + C_2)G}{\omega_1 C_1 C_2}\left(\dfrac{s}{\omega_1}\right) + 1} \qquad (7.22)$$

where

$$\omega_1^2 = b = \frac{g_m G}{\omega_1 C_1 C_2} \qquad (7.23)$$

and

$$a = \frac{(C_1 + C_2)G}{\omega_1 C_1 C_2} \qquad (7.24)$$

Design Equations
For $C_{1n} = C_{2n} = C_n = 1\,F$ and $\omega_1 = 1\,rad/s$ $\therefore$
  From Equation (7.24) we have:

$$R_n = \frac{2}{a} \tag{7.25}$$

and from Equation (7.23) we have:

$$g_{mn} = \frac{2b}{a} \tag{7.26}$$

EXAMPLE 7.1

Design a second-order Butterworth low-pass filter with a cutoff frequency of 100 kHz.

Solution

From Butterworth coefficients we find $a = 1.414$ and $b = 1.000$, hence

$$R_n = \frac{2}{a} = \frac{2}{1.414} = 1.414\,\Omega$$

$$g_{mn} = \frac{2b}{a} = \frac{2}{1.414} = 1.414\,S$$

$$ISF = 10^4 \quad \therefore \quad FSF = \frac{\omega_1}{\omega_n} = 2\pi f_1 = 2\pi \times 10^5 \quad \therefore$$

$$k = ISF \times FSF = 2\pi \times 10^9 \quad \therefore$$

$$C = \frac{C_n}{k} = \frac{1}{2\pi \times 10^9} = 159.2\,pF$$

$$g_m = \frac{g_{mn}}{ISF} = \frac{1.414}{10^4} = 141.4\,\mu S$$

$$R = ISF \times R_n = 10^4 \times 1.414\,\Omega = 14.14\,k\Omega$$

Figure 7.6 shows the designed filter.
  The sensitivity of this filter is

$$S_{g_m}^{1/a} = -S_{1/R}^{1/a} = \frac{1}{2} \tag{7.27}$$

$$S_C^{1/a} = 0 \tag{7.28}$$

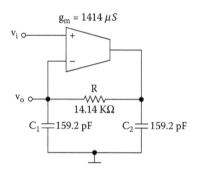

FIGURE 7.6 Second-order Butterworth LPF $f_1 = 100$ kHz.

These results indicate extremely low-sensitivity performance of this OTA filter. This is generally true for other OTA filters.

7.4 SECOND-ORDER LP FILTER WITH FOUR PASSIVE COMPONENTS

Figure 7.7a shows a second-order low-pass filter with four passive components, and the equivalent circuit is shown in Figure 7.7b.

From the equivalent circuit we have:

node v_o

$$(sC_1 + G_1)V_o - GV_a = 0 \tag{7.29}$$

node v_a

$$-G_1V_o + (sC_2 + G_1 + G_2)V_a - g_m(V_i - V_o) = 0 \tag{7.30}$$

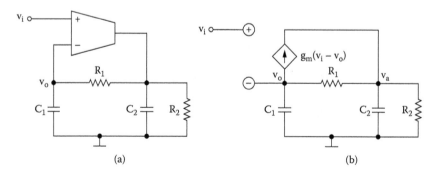

(a)   (b)

FIGURE 7.7 (a) Second-order low-pass filter with four passive components, (b) equivalent circuit.

From the aforementioned equations we have:

$$[(sC_1+G_1)(sC_2+G_1+G_2)-G_1(G_1-g_m)]V_o = g_mG_1V_i \quad \therefore$$

$$H(s) = \frac{V_o}{V_i} = \frac{g_mG_1}{s^2C_1C_2 + s[C_1(G_1+G_2)+C_2G_1]+G_1(g_m+G_2)} \quad \therefore$$

$$H(s) = \frac{K}{\left(\dfrac{s}{\omega_1}\right)^2 + a\left(\dfrac{s}{\omega_1}\right)+1} \tag{7.31}$$

where

$$K = \frac{g_mR_2}{1+g_mR_2} \tag{7.32}$$

$$\omega_1^2 = b = \frac{(g_m+G_2)G_1}{C_1C_2} \tag{7.33}$$

and

$$a = \frac{C_1(G_1+G_2)+C_2G_1}{C_1C_2\omega_1} \tag{7.34}$$

If $C_1 = C_2 = C$ and $R_1 = R_2 = R$, we have:

$$K = \frac{g_mR}{1+g_mR} \tag{7.35}$$

$$\omega_1^2 = b = \frac{(g_m+G)G}{C^2} \tag{7.36}$$

$$a = \frac{3G}{C\omega_1} \tag{7.37}$$

Design Equations
For normalized filter, $\omega_1 = 1\,\text{rad/s}$ and $C_n = 1\,\text{F}$, we have:

$$R_n = \frac{3}{a} \tag{7.38}$$

$$\frac{g_{mn} + \dfrac{1}{R_n}}{R_n} = b \qquad \therefore$$

$$g_{mn} = \frac{9b - a^2}{3a} \qquad\qquad (7.39)$$

EXAMPLE 7.2

Design a second-order Butterworth low-pass filter with $f_1 = 100$ kHz.

Solution

From Butterworth coefficients we find $a = 1.414$, $b = 1.000$.

$$R_n = \frac{3}{a} = \frac{3}{1.414} = 2.122\ \Omega$$

$$g_{mn} = \frac{9b - a^2}{3a} = \frac{9 \times 1 - 1.414^2}{3 \times 1.414} = 1.650\ \text{S}$$

$$ISF = 10^4 \quad \therefore \quad FSF = \frac{\omega_1}{\omega_n} = 2\pi f_1 = 2\pi \times 10^5 \qquad \therefore$$

$$k = ISF \times FSF = 2\pi \times 10^9 \qquad \therefore$$

$$C = \frac{C_n}{k} = \frac{1}{2\pi \times 10^9} = 159.2\ \text{pF},$$

$$R = ISF \times R_n = 10^4 \times 2.122\ \Omega = 21.2\ \text{k}\Omega$$

$$g_m = \frac{g_{mn}}{ISF} = \frac{1.650}{10^4} = 165\ \mu\text{S}$$

$$K = \frac{g_m R}{1 + g_m R} = \frac{165 \times 10^{-6} \times 21.2 \times 10^3}{1 + 165 \times 10^{-6} \times 21.2 \times 10^3} = 0.779$$

Figure 7.8 shows the designed filter.
    The sensitivity of this filter is

$$S_C^{\omega_1} = -S_{1/R}^{\omega_1} = -\frac{1}{2} \qquad\qquad (7.40)$$

$$S_{g_m}^{\omega_1} = \frac{1}{2}\left[1 - \left(\frac{a}{3}\right)^2\right] \qquad\qquad (7.41)$$

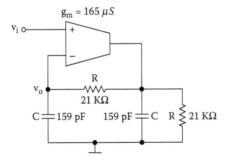

FIGURE 7.8 Second-order Butterworth LPF, with $f_1 = 100$ kHz.

$$S_C^K = 0 \tag{7.42}$$

$$S_{g_m}^K = \left(\frac{a}{3}\right)^2 \tag{7.43}$$

The filter has a very low-sensitivity performance.

7.5 SECOND-ORDER BAND-PASS FILTER

Figure 7.9a shows a band-pass filter, and the equivalent circuit is shown in Figure 7.9b.
From Figure 7.9b we have:
node v_o

$$(2sC + G)V_o - sCV_a = 0 \tag{7.44}$$

node v_a

$$-sCV_o + (sC + G)V_a - g_m(V_i - V_o) = 0 \tag{7.45}$$

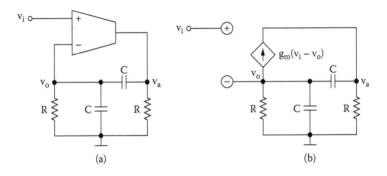

(a)   (b)

FIGURE 7.9 (a) Second-order band-pass filter, (b) equivalent circuit.

From these equations, we take

$$\left[\frac{(sC+G)(2sC+G)}{sC} - sC + g_m \right] V_o = g_m V_i \qquad \therefore$$

$$H(s) = \frac{V_o}{V_i} = \frac{sC g_m}{s^2 C^2 + sC(3G + g_m) + G^2} \qquad \therefore$$

$$H(s) = \frac{K_o \left(\dfrac{s}{\omega_o} \right)}{\left(\dfrac{s}{\omega_o} \right)^2 + \dfrac{1}{Q} \left(\dfrac{s}{\omega_o} \right) + 1} \qquad (7.46)$$

where

$$\omega_o = \frac{G}{C} \qquad (7.47)$$

$$\frac{1}{Q} = \frac{3G + g_m}{\omega_o C} \qquad (7.48)$$

$$K_o = \frac{g_m}{\omega_o C} \qquad (7.49)$$

For $s = j\omega_o \qquad \therefore$

$$H(j\omega_o) = K = \frac{Kj}{j\dfrac{1}{Q}} = K_o Q \qquad \therefore$$

$$K_o = \frac{K}{Q} \qquad (7.50)$$

Design Equations
For $\omega_o = 1$ rad/s, $C_n = 1$ F $\qquad \therefore$

$$R_n = 1\,\Omega \qquad (7.51)$$

From Equation (7.48), we have:

$$\frac{1}{Q} = 3 + g_m \qquad \therefore$$

$$g_{mn} = \frac{3Q - 1}{Q} \qquad (7.52)$$

and from Equation (7.49)

$$K_o = g_{mn} \tag{7.53}$$

$$K = Q g_{mn} \tag{7.54}$$

EXAMPLE 7.3

Design a narrow-band band-pass filter with $f_o = 100$ kHz and $Q = 10$.

Solution

$$g_{mn} = \frac{3Q-1}{Q} = \frac{3 \times 10 - 1}{10} = 2.9 \text{ S}$$

$$C_n = 1 \text{ F} \quad \text{and} \quad R_n = 1 \text{ }\Omega$$

$ISF = 10^4 \qquad \therefore$

$$FSF = \frac{\omega_o}{\omega_n} = 2\pi f_o = 2\pi \times 10^5 \qquad \therefore$$

$$C = \frac{C_n}{ISF \times FSF} = \frac{1}{2\pi \times 10^9} = 159 \text{ pF}$$

$$R = ISF \times R_n = 10^4 \times 1 \text{ }\Omega = 10 \text{ k}\Omega$$

$$g_m = \frac{g_{mn}}{ISF} = \frac{2.9}{10^4} = 290 \text{ }\mu\text{S}$$

Figure 7.10 shows the designed filter.

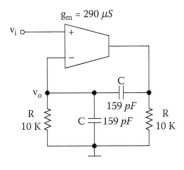

FIGURE 7.10 Narrow-band band-pass filter with $f_o = 100$ kHz and $Q = 10$.

7.6 OTA-C FILTER

A popular OTA application is the realization of fully integrated continuous-time filters. OTA-based filters are referred to as $g_m - C$ filters because they use OTAs and capacitors. Figure 7.11 shows a popular $g_m - C$ filter.

Its analysis proceeds as follows:

$$V_o = V_1^- = V_3^- \tag{7.55}$$

$$V_a = V_1^+ \tag{7.56}$$

where v_1, v_2, and v_3 refer to the input pins of the respective OTAs.

The current out of OTA1 is

$$I_{o1} = g_{m1}(V_1^+ - V_1^-) = g_{m1}(V_a - V_o) \tag{7.57}$$

For OTA2:

$$V_2^+ = I_{o1}\frac{1}{sC_1} = g_{m1}\frac{V_a - V_o}{sC_1} \tag{7.58}$$

Because $V_2^- = 0$, the current out of OTA2 is

$$I_{o2} = g_{m2}(V_2^+ - V_2^-) = \frac{g_{m1}g_{m2}(V_a - V_o)}{sC_1} \tag{7.59}$$

For OTA3, we have:

$$V_3^+ = V_b \quad \text{and} \quad V_3^- = V_o = \frac{I_{o2}}{sC_2} + V_c \tag{7.60}$$

$$I_{o3} = g_{m3}(V_3^+ - V_3^-) = g_{m3}(V_b - V_o) \tag{7.61}$$

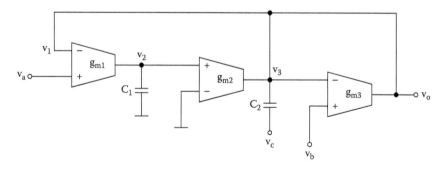

FIGURE 7.11 Second-order g_m-C filter.

At the output of both OTA2 and OTA3, Kirchhoff's current law gives:

$$I_{o2} + I_{o3} = (V_o - V_c)sC_2 \tag{7.62}$$

From Equations (7.59) and (7.62) we have:

$$\frac{g_{m1}g_{m2}(V_a - V_o)}{sC_1} + g_{m3}(V_b - V_o) = (V_o - V_c)sC_2 \quad \therefore$$

$$(s^2C_1C_2 + g_{m3}sC_1 + g_{m1}g_{m2})V_o = s^2C_1C_2V_c + sg_{m3}C_1V_b + g_{m1}g_{m2}V_a \quad \therefore$$

$$V_o = \frac{s^2C_1C_2 + sg_{m3}C_1V_b + g_{m1}g_{m2}V_a}{s^2C_1C_2 + sg_{m3} + g_{m1}g_{m2}} \tag{7.63}$$

For $V_c = 0$, $V_a = 0$, and $V_b = V_i$ $\quad \therefore$

$$H_{BP}(s) = \frac{V_o}{V_i} = \frac{sg_mC_1}{s^2C_1C_2 + sg_{m3}C_1 + g_{m1}g_{m2}} \quad \therefore$$

$$H_{BP}(s) = \frac{\dfrac{1}{Q}\left(\dfrac{s}{\omega_o}\right)}{\left(\dfrac{s}{\omega_o}\right)^2 + \dfrac{1}{Q}\left(\dfrac{s}{\omega_o}\right) + 1} \tag{7.64}$$

where

$$\omega_o^2 = \frac{g_{m1}g_{m2}}{C_1C_2} \tag{7.65}$$

$$Q = \frac{\omega_o C_2}{g_{m3}} \tag{7.66}$$

By setting $g_{m1} = g_{m2} = g_m$, it is clear from Equation (7.65) that the center frequency can be literally dependent on g_m.

$$\omega_o = \frac{g_m}{\sqrt{C_1C_2}} \quad \text{or}$$

$$f_o = \frac{g_m}{2\pi\sqrt{C_1C_2}} \tag{7.67}$$

At the same time, g_{m3} can be *separately* adjusted to yield a controllable value for the filter Q:

$$Q = \frac{1}{g_{m3}} C_2 \frac{g_m}{\sqrt{C_1 C_2}} \qquad \therefore$$

$$Q = \frac{g_m}{g_{m3}} \sqrt{\frac{C_2}{C_1}} \qquad (7.68)$$

Therefore, the band-pass filter realized with this circuit has independently controlled f_o and Q.

If $V_b = V_c = 0$ and $V_a = V_i$ $\quad \therefore$

$$H_{LP}(s) = \frac{V_o}{V_i} = \frac{g_{m1} g_{m2}}{s^2 C_1 C_2 + g_{m3} C_1 s + g_{m1} g_{m2}} \qquad \therefore$$

$$H_{LP}(s) = \frac{1}{\left(\dfrac{s}{\omega_1}\right)^2 + \dfrac{g_{m3}}{\omega_1 C_2}\left(\dfrac{s}{\omega_1}\right) + 1} \qquad (7.69)$$

where

$$\omega_1^2 = b = \frac{g_{m1} g_{m2}}{C_1 C_2} \qquad (7.70)$$

$$a = \frac{g_{m3}}{C_2} \qquad (7.71)$$

For $g_{m1} = g_{m2} = g_m$ $\quad \therefore$

$$C_2 = \frac{g_{m3}}{a} \qquad (7.72)$$

and

$$C_1 = \frac{a\, g_m^2}{b\, g_{m3}} \qquad (7.73)$$

If $V_b = V_a = 0$ and $V_c = V_i$ $\quad \therefore$

$$H_{HP}(s) = \frac{s^2 C_1 C_2}{s^2 C_1 C_2 + g_{m3} C_1 s + g_{m1} g_{m2}} \qquad \therefore$$

$$H_{HP}(s) = \frac{\left(\dfrac{s}{\omega_2}\right)^2}{\left(\dfrac{s}{\omega_2}\right)^2 + \dfrac{g_{m3}}{\omega_2 C_2}\left(\dfrac{s}{\omega_2}\right) + 1} \tag{7.74}$$

where

$$a = \frac{g_{m3}}{\omega_2 C_2} \tag{7.75}$$

and

$$\omega_2^2 = \frac{g_{m1} g_{m2}}{C_1 C_2} \tag{7.76}$$

For $\omega_2 = 1$ rad/s and $g_{m1} = g_{m2} = g_m$ $\qquad \therefore$

$$C_{1n} = \frac{a\, g_m^2}{b\, g_{m3}} \tag{7.77}$$

$$C_{2n} = \frac{g_{m3}}{a} \tag{7.78}$$

7.7 SOME NONIDEAL FEATURE OF THE OTA

One of the biggest drawbacks of the first versions of the OTA was the limited range of the input differential voltage swing. The limited input voltage swing applies only if the OTA is being used in the open-loop configuration. In that case, if the difference-mode voltage exceeds about 25 mV and the load resistance is relatively low, then the circuit does not operate in the linear region. Of course, for the circuits that use negative feedback, the linear behavior is maintained.

The more recent versions of the OTA, such as the National Semiconductor's LM3600, Harris' CA3280A, and Philips' NE5517, all use internal linearizing diodes as the input differential pair of the OTA. These make the OTA's output current a linear function of the amplifier bias current over a wide range of differential input voltage.

As with all active devices, OTAs have finite input and output impedances and a frequency-dependent gain parameter, leading to the more realistic circuit model of Figure 7.12. The capacitors C_i and C_o are not only the parasitic device input and

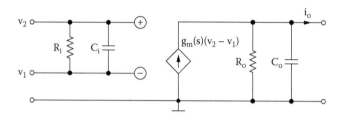

FIGURE 7.12 Practical OTA circuit model.

output capacitances but also those contributed by wiring. The input resistor R_i is from $>10\,MW$ up to $>100\,MW$ (in CMOS circuits), so that its effect can be neglected. The output R_o is usually small, of the order of 100 KW or less, so that its effect must be included.

The transconductance frequency dependence can be modeled via a dominant pole or excess phase shift:

$$g_m(s) = \frac{g_{mo}}{1 + \dfrac{s}{\omega_b}}$$ (7.79)

where ω_b is the bandwidth of the OTA and g_{mo} is the dc transconductance. The phase shift model is also often used, which is

$$g_m(j\omega) = g_{mo}e^{-j\phi} \cong g_{mo}(1 - s\tau)$$ (7.80)

where ϕ is the phase delay, $\tau = 1/\omega_b$ is the time delay, and $\phi = \omega\tau$, when $\omega \ll \omega_b$. Some typical values of OTA (CMOS) parameters are:

$$g_{mo} = 50\,\mu S, \quad f_b = 100\,MHz, \quad R_i = \infty$$

$$R_o = 1\,M\Omega, \quad C_i = 0.05\,pF, \quad \text{and } C_o = 0.1\,pF$$

PROBLEMS

7.1 For the circuit of Figure P7.1, find the input impedance.

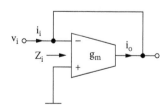

7.2 For the circuit of Figure P7.2, prove that

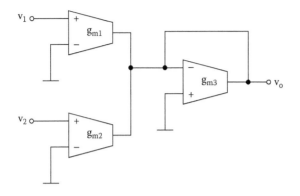

$$V_o = \frac{g_{m1}}{g_{m3}} V_1 + \frac{g_{m2}}{g_{m3}} V_2$$

7.3 Figure P7.3 shows an integrator. Prove that

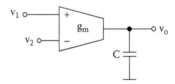

$$v_o = \frac{g_m}{C} \int_0^t (v_2 - v_1)\,dt$$

7.4 Find the transfer function of the filter of Figure P7.4.

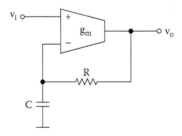

Ans.

$$H(s) = \frac{g_m (sRC + 1)}{sC + g_m}$$

7.5 Prove that the transfer function of the filter of Figure P7.5 is

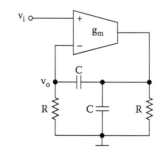

$$H(s) = \frac{V_o}{V_i} = \frac{K_o\left(\dfrac{s}{\omega_o}\right)}{\left(\dfrac{s}{\omega_o}\right)^2 + \dfrac{1}{Q}\left(\dfrac{s}{\omega_o}\right) = 1}$$

where

$$\omega_o = \frac{1}{RC}, \quad Q = \frac{\omega_o RC}{3 + g_m R} \quad \text{and} \quad K_o = \frac{1}{Q}$$

7.6 Find the transfer function of Figure P7.6.

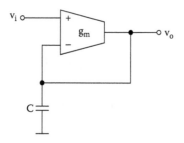

Ans.

$$H(s) = \frac{g_m}{sC + g_m}$$

7.7 Find the transfer function of Figure P7.7.

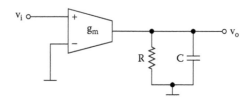

Ans.

$$H(s) = \frac{g_m}{sC + G}$$

7.8 Find the transfer functions of Figure P7.8.

(a) $H(s) = \dfrac{V_o}{V_i}$, (b) $H(s) = \dfrac{V_a}{V_i}$

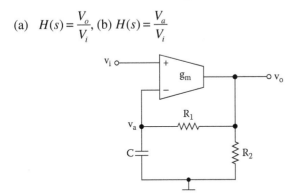

Ans.

(a) $H(s) = \dfrac{sg_m C + g_m G_1}{sC(G_1 + G_2) + G_1(g_m + G_2)}$

(b) $H(s) = \dfrac{g_m G_1}{sC(G_1 + G_2) + G_1(g_m + G_2)}$

7.9 Find $v_o = f(v_1, v_2)$ of the circuit of Figure P7.9.

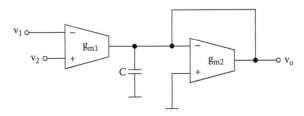

Ans.

$$Vo = \frac{g_{m1}}{sC + g_{m2}}(V_2 - V_1)$$

7.10 Find the transfer function of Figure P7.10.

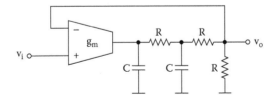

Ans.

$$H(s) = \frac{K}{\left(\dfrac{s}{\omega_1}\right)^2 + \dfrac{1}{Q}\left(\dfrac{s}{\omega_1}\right) + 1}, \quad K = \frac{g_m}{g_m + G}, \quad \omega_1^2 = \frac{G(g_m + G)}{2C^2}$$

$$Q = \frac{2\omega_1 RC}{5}$$

7.11 Find the transfer function of the $g_m - C$ filter of Figure P7.11.

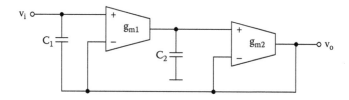

8 Switched Capacitor Filters

8.1 INTRODUCTION

Many active filters with resistors and capacitors have been replaced with a filter called a *switched capacitor filter* (SC filter). The switched capacitor filter allows for very sophisticated, accurate, and tunable analog circuits to be manufactured without using resistors. The main reason is that resistors are hard to build on integrated circuits because they take up a lot of space, and the circuits can be made to depend on ratios of capacitor values, which can be set accurately, and not absolute values, which vary between manufacturing runs.

8.2 THE SWITCHED CAPACITOR RESISTORS

To understand how a switched capacitor circuit works, consider Figure 8.1a, in which the capacitor is connected to two switches and two different voltages. Assuming $V_2 \gg V_1$, we observe that flipping the switch to the left changes C to V_1, and flipping it to the right discharges C to V_2. The net charge transfer from V_1 to V_2 is

$$\Delta q = C(V_2 - V) \tag{8.1}$$

If the switched process is repeated n times in a time Δt is given by

$$\Delta q = C(V_2 - V)\frac{n}{\Delta t} \tag{8.2}$$

The left-hand side of this equation represents current, and the number of cycles per unit time is the switching frequency or clock frequency, f_{CLK}, hence:

$$i = C(V_2 - V)f_{CLK}$$

Rearranging, we get

$$R_{eq} = \frac{V_2 - V_1}{i} = \frac{1}{Cf_{CLK}} \tag{8.3}$$

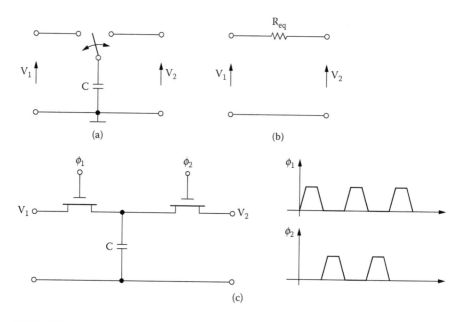

FIGURE 8.1 (a) Switched capacitor; (b) approximate resistor; (c) MOS implementation; clock signals ϕ_1 and ϕ_2.

This equation indicates that the *switched capacitor* behaves approximately like a resistor. The value of the resistor decreases with increasing switching frequency or increasing capacitance.

8.3   THE SWITCHED CAPACITOR INTEGRATOR

An active RC integrator is shown in Figure 8.2a. The input resistance to the integrator, R, is changed to a switched capacitor in the circuit of Figure 8.2b.

The input capacitor C_1 charges to V_i during the first half of the clock period; that is,

$$q = C_1 V_i$$

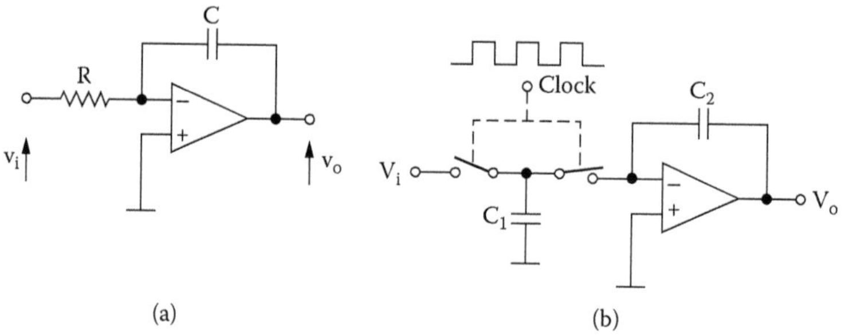

FIGURE 8.2 (a) Active integrator; (b) switched-capacitor integrator.

Because the clock frequency f_{CLK} is much higher than the frequency being filtered, V_i does not change, while C_1 is being changed.

During the second half of the clock period, the charge C_1V_i is transferred to the feedback capacitor because $v^- = v^+ = 0$. The total transfer of charge in one clock cycle is

$$q = C_1V_i \qquad (8.4)$$

Hence, the average input current is

$$i_i = \frac{q}{T} = \frac{C_1V_i}{T} = C_1V_i f_{CLK} \qquad (8.5)$$

The equivalent input resistor (R_{eq}) can be expressed as

$$R = \frac{V_i}{i_i} = \frac{1}{C_1 f_{CLK}} \qquad (8.6)$$

and the equivalent RC time constant for the SC filter is

$$RC = \frac{C_2}{C_1 f_{CLK}} \qquad (8.7)$$

The RC time constant that determines the integrator frequency response is determined by the clock frequency, f_{CLK}, and the capacitor ratio, C_2 / C_1. The clock frequency can be set with an accurate crystal oscillator. The capacitor ratios are accurately fabricated on an IC chip with a typical tolerance of 0.1%. We need not accurately set a value of C_2, because only the ratio C_2 / C_1 affects the time constant. We use small values of capacitances, such as 0.1 pF, to reduce the area on the IC devoted to the capacitors. We can obtain relative large time constants [Equation (8.7)], suitable for audio applications, with small areas on the IC.

The frequency of the active integrator is given by

$$f_1 = \frac{1}{2\pi RC}$$

Replacing R by the SR resistance, we have:

$$f_1 = \frac{C_1}{2\pi C_2} f_{CLK} \qquad (8.8)$$

The input frequency f is such that

$$f \ll f_{CLK} \qquad (8.9)$$

8.4 UNIVERSAL SC FILTERS

The universal SC filter consists of two high-performance, SC filters. Each filter, together with two to five external resistors, can produce various second-order filter functions such as low-pass, high-pass, band-pass, notch, and all pass. The center frequency of these functions can be tuned by an external clock. Up to fourth-order full biquadratic functions can be achieved by cascading the two filter blocks. Any of the classical filter configurations, such as Butterworth, Chebyshev, Bessel, and Cauer, can be formed. Two popular and well-documented examples are the MF10 or MF100 (National Semiconductor) and the LTC 1060 (Linear Technology).

8.4.1 THE LMF100 UNIVERSAL SC FILTER

The MF100 or LMF100 integrated circuit is a versatile circuit with four SC integrators that can be connected as two second-order filters or one fourth-order filter. This chip includes a pin that selects the clock to center frequency ratio at either 50:1 and 100:1. That is,

$$f_c = \frac{f_{CLK}}{50} \quad \text{or} \quad \frac{f_{CLK}}{100} \tag{8.10}$$

where f_c is the *integration unity-gain frequency* depending on the state of the $50/100$ pin. Figure 8.3 shows the block diagram of the left half.

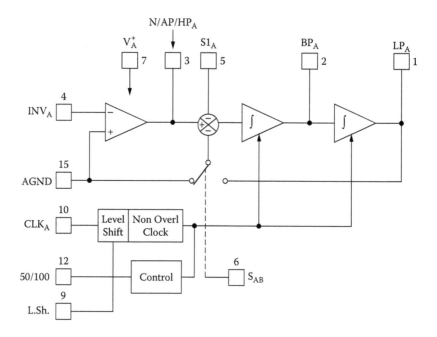

FIGURE 8.3 Half-block diagram of the MF100 Universal monolithic dual SC filter.

The pins are described in the data sheet, and we will describe a few of them:

- 50/100: determines if the value of f_c is $f_{CLK} / 100$, or $f_{CLK} / 50$.
- CLK_A: is f_{CLK}.
- INV_A: is inverting input of the op-amp.
- $N / AP / HP_A$: an intermediate output, and the noninverting input to the summer. Used for notch (N), all-pass (AP), or high-pass (HP) output.
- BP_A: the output of the first integrator. Used for band-pass (BP) output.
- LP_A: the output of the second integrator. Used for low-pass (LP) output.
- $S1_A$: an inverting input to the summer.
- S_{AB}: determines if the switch is to left or to the right, i.e., this pin determines if the second inverting input to the summer is ground (AGND), or the low-pass output.

8.4.1.1 Modes of Operation

Each section can be configured for a variety of different modes. The mode of Figure 8.4 is referred to as the *state variable* mode (mode 3) because it provides the high-pass, band-pass, and low-pass responses by direct consecutive integrations; other modes can be found in the data sheets and application notes.
node v_a

$$-G_1 V_i + (G_1 + G_2 + G_3 + G_4) V_a - G_2 V_{HP} - G_3 V_{BP} - G_4 V_{LP} = 0 \qquad (8.11)$$

$$V_{BP} = \frac{1}{s} V_{HP} \qquad (8.12)$$

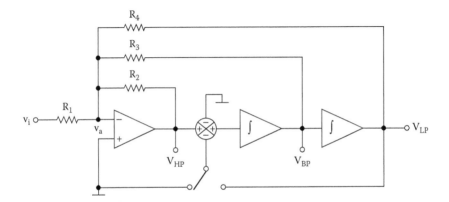

FIGURE 8.4 Mode 3 (HPF, BPF, and LPF) of SC filter MF100.

$$V_{LP} = \frac{1}{s} V_{BP} \tag{8.13}$$

$$V_a = 0 \text{ (for an ideal op-amp)} \tag{8.14}$$

For Equations (8.11) and (8.14), we have:

$$G_4 V_{LP} + G_2 V_{HP} + G_3 V_{BP} = -G_1 V_i \tag{8.15}$$

8.4.1.2 Low-Pass Filter

From Equations (8.15), (8.12), and (8.13) we have:

$$(s^2 G_2 + s G_3 + G_4) V_{LP} = -G_1 V_i \quad \therefore$$

$$H_{LP}(s) = \frac{V_{LP}}{V_i} = -\frac{G_1}{s^2 G_2 + s G_3 + G_4} \quad \therefore$$

$$H_{LP}(s) = -\frac{\dfrac{R_2}{R_1}}{s^2 + \dfrac{R_2}{R_3} s + \dfrac{R_2}{R_4}}$$

For $s \to 0$ $\quad \therefore$

$$H_{LP}(0) = K = \frac{R_4}{R_1} \tag{8.16}$$

$\therefore$

$$H_{LP}(s) = -\frac{\dfrac{R_2}{R_1}}{s^2 + a s + b} \tag{8.17}$$

where

$$a = \frac{R_2}{R_3} \tag{8.18}$$

$$\omega_1^2 = b = \frac{R_2}{R_4} \tag{8.19}$$

For a normalized low-pass filter, we have:

$$R_{1n} = 1\,\Omega, \quad \omega_{1n} = 1 \text{ rad/s} \tag{8.20}$$

$$R_{4n} = K R_{1n} = K\,\Omega \tag{8.21}$$

$$R_{2n} = bK\,\Omega \tag{8.22}$$

$$R_{3n} = \frac{R_{2n}}{a} = \frac{bK}{a}\,\Omega \tag{8.23}$$

EXAMPLE 8.1

Design a low-pass Butterworth filter with the following specifications: $f_1 = 1\text{ kHz}$, $f_s = 2.5\text{ kHz}$, $A_{max} = 3\text{ dB}$, and $K = 1$.

Solution

From the Butterworth nomographs, we find: $n = 5$, and from Butterworth coefficients, we have:

$$a_1 = 1.618,\ b_1 = 1.000,\ a_2 = 0.617,\ b_2 = 1.000,\ \text{and}\ b_3 = 1.000$$

For $K = 1$ we have:

$$R_{1n} = R_{2n} = R_{4n} = 1\,\Omega$$

First stage

$$R_{3n} = \frac{R_{2n}}{a} = \frac{1}{1.618} = 0.617\,\Omega$$

Second stage

$$R_{3n} = \frac{R_{2n}}{a} = \frac{1}{0.618} = 1.618\,\Omega$$

Third stage

$$R_n = 1\,\Omega$$

$$C_n = \frac{1}{b} = 1\text{ F}$$

$ISF = 2 \times 10^4$; we calculate

$$FSF = \frac{\omega_1}{\omega_n} = 2\pi f_1 = 2\pi \times 10^3$$

$$R_1 = R_2 = R_4 = ISF \times R_{1n} = 2 \times 10^4 \times 1\,\Omega = 20\text{ k}\Omega$$

First stage

$$R_3 = ISF \times R_{3n} = 2 \times 10^4 \times 0.617\,\Omega = 12.3\text{ k}\Omega$$

Second stage

$$R_3 = ISF \times R_{3n} = 2 \times 10^4 \times 1.618\,\Omega = 32.4\text{ k}\Omega$$

Third stage

$$R = ISF \times R_n = 2 \times 10^4 \times 1\,\Omega = 20\text{ k}\Omega$$

$$C = \frac{C_n}{ISF \times FSF} = \frac{1}{2 \times 10^4 \times 2\pi \times 10^3} \cong 8\text{ nF}$$

$$f_{CLK} = 50\,f_1 = 50 \times 1\text{ kHz} = 50\text{ kHz}$$

Figure 8.5 shows the designed filter using MF5 (half of MF100).

8.4.1.3 High-Pass Filter

From Equations (8.15) and (8.13), we take:

$$G_4\left(\frac{1}{s}V_{BP}\right) + G_2 V_{HP} + G_3\left(\frac{1}{s}V_{HP}\right) = -G_1 V_i \qquad \therefore$$

$$G_4\left(\frac{1}{s}\frac{1}{s}V_{HP}\right) + G_2 V_{HP} + \frac{G_3}{s}V_{HP} = -G_1 V_i \qquad \therefore$$

$$H_{HP}(s) = \frac{V_{HP}}{V_i} = -\frac{s^2 G_1}{s^2 G_2 + s G_3 + G_4} \qquad \therefore$$

$$H_{HP}(s) = -\frac{\dfrac{R_2}{R_1}s^2}{s^2 + \dfrac{R_2}{R_3}s + \dfrac{R_2}{R_4}} \qquad (8.24)$$

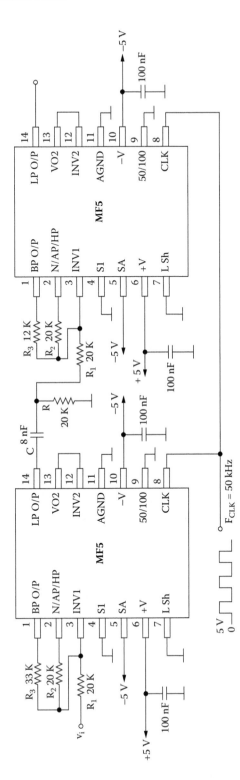

FIGURE 8.5 Fifth-order SC LP Butterworth filter: $f_1 = 1$ kHz, $K = 1$.

For $s \to \infty$ ∴

$$H_{HP}(\infty) = K = \frac{R_2}{R_1} \qquad (8.25)$$

$$H_{HP}(s) = -\frac{K}{s^2 + as + b} \qquad (8.26)$$

where

$$\omega_2^2 = b = \frac{R_2}{R_4} \qquad (8.27)$$

$$a = \frac{R_2}{R_3} \qquad (8.28)$$

For $\omega_{2n} = 1$ rad/s and $R_{1n} = 1\,\Omega$ ∴

$$R_{2n} = K \qquad (8.29)$$

$$R_{3n} = \frac{1}{a} \qquad (8.30)$$

$$R_{4n} = \frac{K}{b} \qquad (8.31)$$

EXAMPLE 8.2

Design an SC high-pass Butterworth filter with the following specifications:

$$f_2 = 430\,\text{Hz}, \; f_s = 215\,\text{Hz}, \; A_{max} = 3\,\text{dB}, \; A_{min} = 21\,\text{dB}, \; \text{and} \; K = 1$$

Solution

From the Butterworth nomographs, we find $n = 4$.
First stage

$$a = 1.848, \quad b = 1.000$$

$$R_{1n} = 1\,\Omega, \quad R_{2n} = K = 1\,\Omega, \quad R_{3n} = \frac{1}{a} = \frac{1}{1.848} = 0.541\,\Omega$$

$$R_{4n} = \frac{K}{b} = \frac{1}{1} = 1\,\Omega$$

Second stage

$$a = 0.766, \quad b = 1.000$$

$$R_{1n} = 1\,\Omega, \quad R_{2n} = K = 1\,\Omega$$

$$R_{3n} = \frac{1}{a} = \frac{1}{0.766} = 1.305\,\Omega$$

$$R_{4n} = \frac{K}{b} = 1\,\Omega$$

$ISF = 2 \times 10^4 \qquad \therefore$

$$FSF = \frac{\omega_1}{\omega_n} = 2\pi f_1 = 2\pi \times 430 = 8.6\pi \times 10^2$$

$$R_1 = R_2 = R_4 = ISF \times R_{1n} = 2 \times 10^4\,\Omega = 20\text{ k}\Omega$$

First stage

$$R_3 = ISF \times R_{3n} = 2 \times 10^4 \times 0.541\,\Omega = 10.8\text{ k}\Omega$$

Second stage

$$R_3 = ISF \times R_{3n} = 2 \times 10^4 \times 1.305\,\Omega = 26.1\text{ k}\Omega$$

Figure 8.6 shows the designed filter.

8.4.1.4 Narrow-Band Band-Pass Filter

From Equations (8.15), (8.13), and (8.14), we have:

$$G_4\left(\frac{1}{s}V_{BP}\right) + G_2(s\,V_{BP}) + G_3 V_{BP} = -G_1 V_i \qquad \therefore$$

$$H_{BP}(s) = \frac{V_{BP}}{V_i} = -\frac{\dfrac{R_2}{R_1}}{s^2 + \dfrac{R_2}{R_3}s + \dfrac{R_2}{R_4}} \qquad \therefore$$

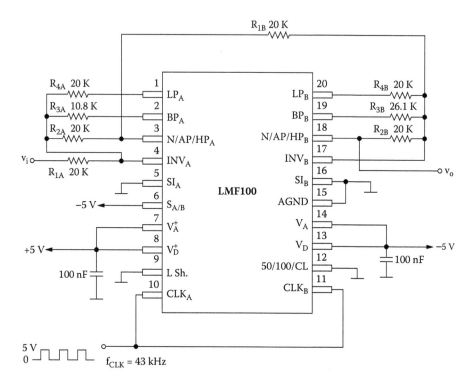

FIGURE 8.6 Fourth-order SC HP Butterworth filter: $f_2 = 430$ Hz, $K = 1$.

$$H_{BP}(s) = -\frac{\dfrac{R_2}{\omega_o R_1}\left(\dfrac{s}{\omega_o}\right)}{\left(\dfrac{s}{\omega}\right)^2 + \dfrac{1}{Q}\left(\dfrac{s}{\omega_o}\right) + 1} \tag{8.32}$$

where

$$\omega_o^2 = \frac{R_2}{R_4} \tag{8.33}$$

$$\frac{1}{Q} = \frac{R_2}{\omega_o R_3} \tag{8.34}$$

For $s = j\omega_o$, Equation (8.32) becomes

$$H_{BP}(j\omega_o) = K = -\frac{R_3}{R_1} \tag{8.35}$$

For normalized filter, $\omega_o = 1$ rad/s:

$$R_{1n} = 1\,\Omega \tag{8.36}$$

From Equations (8.35) and (8.33), we have:

$$R_{3n} = K \tag{8.37}$$

$$R_{2n} = R_{4n} = \frac{K}{Q} \tag{8.38}$$

EXAMPLE 8.3

Design a narrow-band band-pass filter with $f_o = 750$ Hz, $Q = 10$, and $K = 5$.

Solution

$$R_{1n} = 1\,\Omega, \quad R_{2n} = \frac{K}{Q} = \frac{5}{10} = 0.5\,\Omega, \quad R_{3n} = K = 5\,\Omega$$

$$R_{4n} = R_{2n} = 0.5\,\Omega$$

$ISF = 2 \times 10^4 \quad \therefore$

$$R_1 = ISF \times R_{1n} = 2 \times 10^4 \times 1\,\Omega = 20\text{ k}\Omega$$

$$R_2 = R_4 = ISF \times R_{2n} = 2 \times 10^4 \times 0.5\,\Omega = 10\text{ k}\Omega$$

$$R_3 = ISF \times R_{3n} = 2 \times 10^4 \times 5\,\Omega = 100\text{ k}\Omega$$

Figure 8.7 shows the designed filter.

EXAMPLE 8.4

Design a Butterworth band-pass filter with the following specifications: $BW = 2$ to 8 kHz, $f_{s1} = 16$ kHz, $f_{s2} = 1$ kHz, $A_{max} = 3$ dB, $A_{min} = 12$ dB, and $K = 9$.

Solution

From the Butterworth nomographs, we find $n = 2$.
(a) *Low-pass filter*

$$a = 1.414, \quad b = 1.000$$

$$K_1 = K_2 = \sqrt{K} = \sqrt{9} = 3$$

$$R_{1n} = 1\,\Omega, \quad R_{2n} = KR_{1n} = 3 \times 1\,\Omega = 3\,\Omega, \quad R_{4n} = 3\,\Omega$$

$$R_{3n} = \frac{3}{a} = \frac{3}{1.414} = 2.121\,\Omega$$

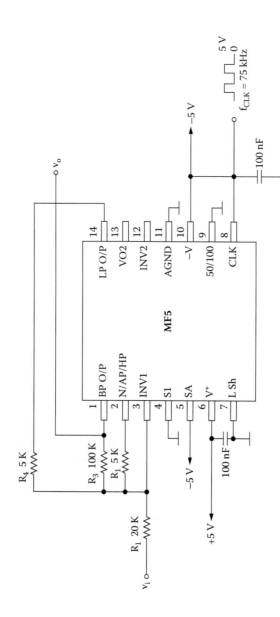

FIGURE 8.7 Narrow-band band-pass filter: $f_o = 750$ Hz, $Q = 10$, $K = 5$.

$ISF = 2 \times 10^4 \qquad \therefore$

$$R_1 = ISF \times R_{1n} = 2 \times 10^4 \times 1\,\Omega = 20\ \text{k}\Omega$$

$$R_2 = R_4 = ISF \times R_{2n} = 2 \times 10^4 \times 3\,\Omega = 60\ \text{k}\Omega$$

$$R_3 = ISF \times R_{3n} = 2 \times 10^4 \times 2.121\,\Omega = 42.4\ \text{k}\Omega$$

(b) *High-pass filter*

$$R_{1n} = 1\,\Omega, \quad R_{2n} = 3R_{1n} = 3\,\Omega, \quad R_{4n} = 3\,\Omega$$

$$R_{3n} = \frac{3}{a} = \frac{3}{1.414} = 2.121\,\Omega$$

$ISF = 2 \times 10^4 \qquad \therefore$

$$R_1 = ISF \times R_{1n} = 2 \times 10^4 \times 1\,\Omega = 20\ \text{k}\Omega$$

$$R_2 = R_4 = ISF \times R_{2n} = 2 \times 10^4 \times 3\,\Omega = 60\ \text{k}\Omega$$

$$R_3 = ISF \times R_{3n} = 2 \times 10^4 \times 2.121\,\Omega = 42.4\ \text{k}\Omega$$

Figure 8.8 shows the designed filter.

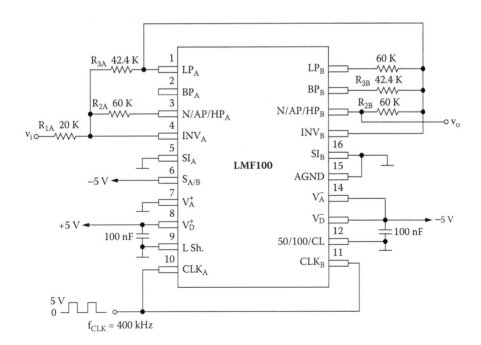

FIGURE 8.8 Broadband band-pass SC filter: $BW = 2$ to 8 kHz, $K = 9$.

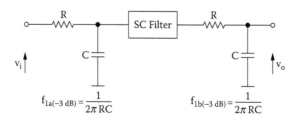

FIGURE 8.9 RC passive antialias filter, f_{1a}, and to remove sampling steps from the output of SC filters, f_{1b}.

8.5 PRACTICAL LIMITATIONS OF SC FILTERS

There are some limitations we must be aware of when using SC filters. These are limits on the permissible range of clock frequency. The permissible clock range is typically between 100 Hz and 1 MHz.

SC filters are sampled data systems, and as such, they possess some characteristics not found in conventional continuous-time filters.

The maximum frequency component a sampled data system can accurately handle is its *Nyquist limit;* that is, *the sample rate must be greater than or equal to twice the highest frequency component in the input signal,* $f_s \geq 2f_h$. When this rule is violated, unwanted or undesirable signals appear in the frequency band of interest. This is called *aliasing.*

For example, to digitize a 3-kHz signal, a minimum sampling frequency of 6 kHz is required. In practice, sampling is usually higher to provide some margin and make the filtering requirements less critical.

The sampling process results in an output signal that changes amplitude every clock period. This creates the stairstep characteristic of a sine wave input. The steps depend entirely on the clock rate and the rate of change of the output voltage.

Because the steps are relatively small compared to the signal amplitude, they are often not bothersome. If they must be reduced, we usually use a simple RC filter at the output of SC filter.

Also, it is a good idea, when very wide-band input signals are encountered, to provide some antialias prefiltering using a single-pole low-pass filter before the SC filter. The prefilter need not be very accurate, as it will be operating at frequencies well above the SC filter's center frequency (Figure 8.9).

PROBLEMS

8.1 The circuit of Figure P8.1 provides the notch, band-pass, and low-pass responses (node 1), with the notch frequency f_z and the response frequency f_o independent prove that

$$K_N = K_{LP} = \frac{R_2}{R_1}, \quad K_{BP} = \frac{R_3}{R_1}, \quad Q = \frac{R_3}{R_2}$$

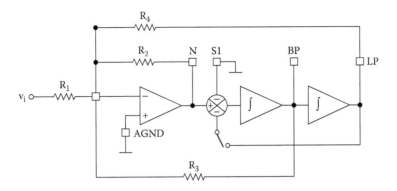

$$f_z = f_o = f_1, \quad f_1 = \frac{f_{CLK}}{100} \quad \text{or} \quad f_1 = \frac{f_{CLK}}{50}$$

8.2 The circuit of Figure P8.2 provides the all-pass and low-pass responses (mode 4). Prove that

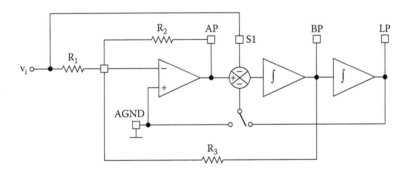

$$Q_p = \frac{R_3}{R_2}, \quad Q_z = \frac{R_3}{R_1}, \quad K = \frac{R_2}{R_1}$$

$$K_{LP} = \left(1 + \frac{R_2}{R_1}\right), \quad K_{BP} = \frac{R_3}{R_2}\left(1 + \frac{R_2}{R_1}\right)$$

8.3 The circuit of Figure P8.3 provides the notch, band-pass, and low-pass responses (mode 1). Prove that

$$K_{LP} = K_N = \frac{R_2}{R_1}, \quad K_{BP} = \frac{R_3}{R_1}, \quad Q = \frac{R_3}{R_2} \quad (f_N = f_o)$$

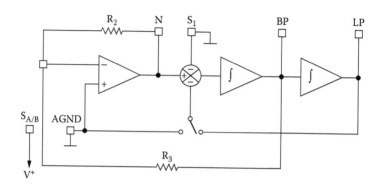

8.4   The circuit of Figure P8.4 provides band-pass and low-pass responses (mode 2). Prove that

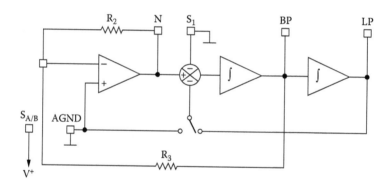

$$f_o = \frac{f_{CLK}}{100}\sqrt{1+\frac{R_2}{R_4}} \quad \text{or} \quad \frac{f_{CLK}}{50}\sqrt{1+\frac{R_2}{R_4}}, \quad f_o : \text{center frequency}$$

$$Q = \frac{\sqrt{1+\dfrac{R_2}{R_4}}}{\dfrac{R_2}{R_4}}, \quad K_{LP} = \frac{\dfrac{R_2}{R_1}}{1+\dfrac{R_2}{R_4}}, \quad K_{BP} = \frac{R_3}{R_1}$$

$$K_{N1} = \frac{\dfrac{R_2}{R_1}}{1+\dfrac{R_2}{R_4}} \ (f \to 0), \quad K_{N2} = \frac{R_2}{R_1}\left(f \to \frac{f_{CLK}}{2}\right)$$

8.5 The circuit of Figure 8.5 provides numerator complex zeros, BP, and LP responses (mode 5). Prove that

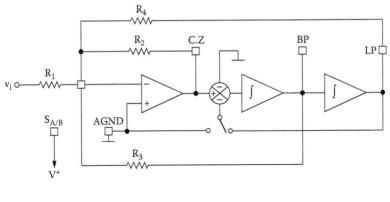

$$f_o = \frac{f_{CLK}}{100}\sqrt{1+\frac{R_2}{R_4}} \quad \text{or} \quad \frac{f_{CLK}}{50}\sqrt{1+\frac{R_2}{R_4}}$$

$$f_o = \frac{f_{CLK}}{100}\sqrt{1+\frac{R_1}{R_4}} \quad \text{or} \quad \frac{f_{CLK}}{50}\sqrt{1+\frac{R_1}{R_4}}$$

$$Q_p = \frac{R_3}{R_2}\sqrt{1+\frac{R_2}{R_4}}, \quad Q_z = \frac{R_3}{R_1}\sqrt{1+\frac{R_1}{R_4}}$$

$$K_{z1} = \frac{R_2(R_4-R_1)}{R_1(R_2+R_4)} \; (f \to 0), \quad K_{z2} = \frac{R_2}{R_1}\left(f \to \frac{f_{CLK}}{2}\right)$$

$$K_{BP} = \frac{R_3}{R_2}\left(1+\frac{R_2}{R_1}\right), \quad K_{LP} = \frac{R_4}{R_1}\left(\frac{R_1+R_2}{R_2+R_4}\right)$$

8.6 The circuit of Figure P8.6 provides single-pole HP and LP filters (mode 6a). Prove that

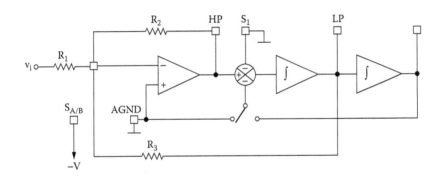

$$f_c = \frac{f_{CLK}}{100}\frac{R_2}{R_3} \quad \text{or} \quad f_c = \frac{f_{CLK}}{50}\frac{R_2}{R_3}, \quad f_c : f_1 \text{ or } f_2$$

$$K_{LP} = -\frac{R_3}{R_1}, \quad K_{HP} = -\frac{R_2}{R_1}$$

8.7 The circuit of Figure P8.7 provides single-pole LP (inverting and noninverting) (mode 6b). Prove that

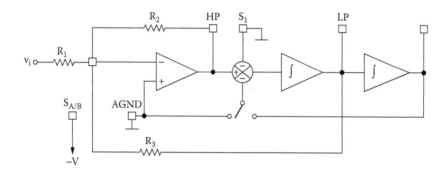

$$f_c = \frac{f_{CLK}}{100}\frac{R_2}{R_3} \quad \text{or} \quad f_c = \frac{f_{CLK}}{50}\frac{R_2}{R_3}$$

$$K_{LP1} = 1 \text{ (noninverting)}, \qquad K_{LP2} = \frac{R_3}{R_2} \text{ (inverting)}$$

8.8 Using the MF100, design a minimum-component fourth-order Butterworth low-pass filter with $f_1 = 2$ kHz and 20-dB dc gain.

8.9 Using MF100, design a fourth-order 1-dB Chebyshev 0.1-dB high-pass filter with $f_2 = 500$ kHz and 0-dB dc gain.

8.10 One of the numerous SC filters produced by National Semiconductor Corporation is the MF6, a sixth-order SC Butterworth low-pass filter (Figure P8.8).

The ratio of the clock frequency to the low-pass cutoff frequency is internally set to 100:1.

Design a Butterworth filter to satisfy the following specifications: $A_{max} = 3$ dB, $f_1 = 3$ kHz, 3 kHz, $A_{min} = 35$ dB, and $f_s = 6$ kHz.

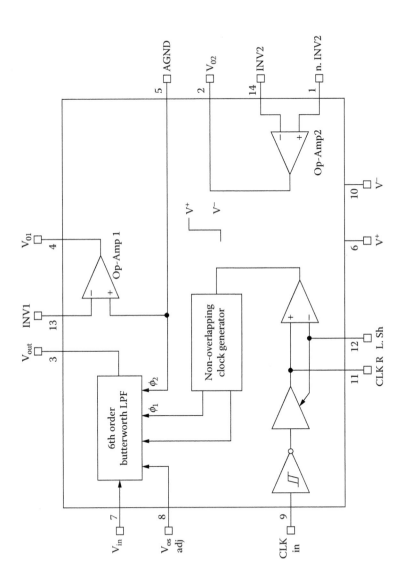

Appendix A: Node Voltage Network Analysis

The node-voltage method of circuit analysis is a method in which the Kirchhoff voltage law (KVL) equations are written implicitly on the circuit diagram so that only the Kirchhoff current law (KCL) equations need to be solved. The method also permits a minimum number of voltage variables to be assigned. The method will be developed through a study of the circuit of Figure A.1. In this circuit, two unknown voltages, v_1 and v_2, are chosen. The voltage v_1 is chosen as a voltage rise from node 3 to node 1; v_2 is similarly chosen as a voltage rise from node 3 to node 2. As node 3 is the point from which the unknown voltages are measured, it is called the **reference node**.

The voltage rise from node 2 to node 1 is the third unknown voltage in the circuit; it is found from the KVL equation to be

$$V_{21} = V_1 - V_2$$

There are three nodes in the circuit, and as a consequence, two independent KCL equations can be written, assuming that all branch currents are leaving the node.

node v_1

$$\frac{V_1 - V_a}{Z_1} + \frac{V_1}{Z_2} + \frac{V_1 - V_2}{Z_3} = 0 \tag{A.1}$$

The selection of the direction of the currents is arbitrary.

node v_2

$$\frac{V_2 - V_1}{Z_3} + \frac{V_2}{Z_4} + \frac{V_2 - V_b}{Z_5} = 0 \tag{A.2}$$

Rearranging the terms of equations (A.1) and (A.2) gives

$$\left(\frac{1}{Z_1} + \frac{1}{Z_2} + \frac{1}{Z_3} \right) V_1 - \frac{1}{Z_3} V_2 = \frac{1}{Z_1} V_a \tag{A.3}$$

$$-\frac{1}{Z_3} V_1 + \left(\frac{1}{Z_3} + \frac{1}{Z_4} + \frac{1}{Z_5} \right) V_2 = \frac{1}{Z_5} V_b \tag{A.4}$$

247

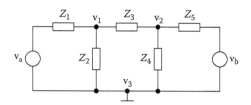

FIGURE A.1 Circuit illustrating the node-voltage method.

or

$$(Y_1 + Y_2 + Y_3)V_1 - Y_3V_2 = Y_1V_a \tag{A.5}$$

$$-Y_3V_1 + (Y_3 + Y_4 + Y_5)V_2 = Y_5V_b \tag{A.6}$$

Examination of Equations (A.5) and (A.6) shows a pattern that will permit equations of this type to be written readily by inspection. In Equation (A.5), written at node 1, the coefficient of v_1 is the positive sum of the admittance connected to node 1. The coefficient of v_2 is the negative sum of the admittances connected between nodes v_1 and v_2. The right-hand side of the equation is the sum of the current sources feeding into node 1.

Now consider Equation (A.6), written for node 2. An analogous situation exists: The coefficient of v_2 is the positive sum of the admittances connected to node 2; the coefficient of v_1 is the negative sum of the admittances between nodes 2 and 1; and the right-hand side of the equation is the sum of the current sources feeding into 2. The fact that these two equations are similar in structure is not a coincidence. It follows from KCL equations and the manner in which the voltage variables are selected.

The formal procedure for writing equations of the type represented by Equations (A.5) and (A.6) is called the **node-voltage method**.

EXAMPLE A.1

Use the node-voltage to determine the transfer function in the circuit of Figure A.2.

$$v_2 = v_0$$

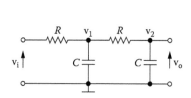

FIGURE A.2 Circuit of Example A.1.

Solution

node v_1

$$-GV_i + (G + sC)V_1 - GV_0 = 0 \qquad \text{(A.7)}$$

node v_2

$$-GV_1 + (sC + G)V_0 = 0 \qquad \text{(A.8)}$$

From Equation (A.8), we have:

$$V_1 = \frac{sC + G}{G} V_0 \qquad \text{(A.9)}$$

From Equations (A.7) and (A.9), we have:

$$\frac{(sC + G)(sc + 2G)}{G} V_0 - GV_0 = GV_i \qquad \therefore$$

$$[(sC + G)(sC + 2G) - G^2]V_0 = G^2 V_i \qquad \therefore$$

$$H(s) = \frac{V_0}{V_i} = \frac{G^2}{s^2 C^2 + 3sCG + G^2} \qquad \therefore$$

$$H(s) = \frac{1}{s^2 R^2 C^2 + 3sRC + 1} \qquad \text{(A.10)}$$

Appendix B: Filter Design Nomograph

B.1 BUTTERWORTH LP FILTER DESIGN NOMOGRAPH

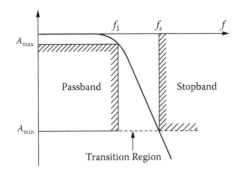

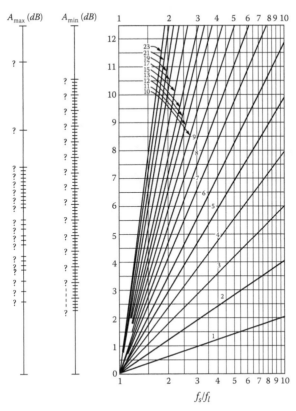

B.2 CHEBYSHEV LP FILTER DESIGN NOMOGRAPH

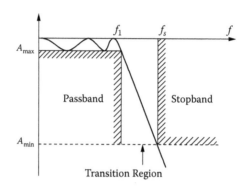

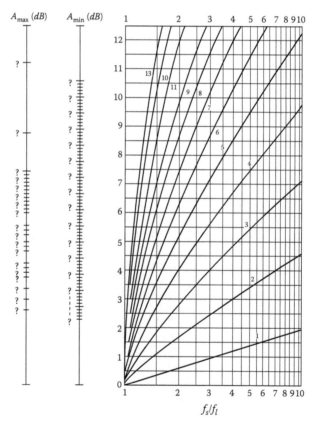

f_s/f_l

Appendix C: First- and Second-Order Factors of Denominator Polynomial

$$D = (s + c_i) \prod_{i=1}^{n} (s^2 + á_i s + b_i)$$

Butterworth

n

1	$(s + 1)$
2	$(s^2 + 1.414s + 1)$
3	$(s + 1)(s^2 + s + 1)$
4	$(s^2 + 1.848s + 1)(s^2 + 0.765s + 1)$
5	$(s + 1)(s^2 + 1.618s + 1)(s^2 + 0.765s + 1)$
6	$(s^2 + 1.932s + 1)(s^2 + 1.414s + 1)(s^2 + 0.518s + 1)$
7	$(s + 1)(s^2 + 1.802s + 1)(s^2 + 1.247s + 1)(s^2 + 0.445s + 1)$
8	$(s^2 + 1.962s + 1)(s^2 + 1.663s + 1)(s^2 + 1.111s + 1)(s^2 + 0.390s + 1)$

Chebyshev 0.1 dB

n

1	$(s + 6.552)$
2	$(s^2 + 2.372s + 3.314)$
3	$(s + 0.969)(s^2 + 0.969s + 1.690)$
4	$(s^2 + 1.275s + 0.623)(s^2 + 0.528s + 1.330)$
5	$(s + 0.539)(s^2 + 0.872s + 0.636)(s^2 + 0.333s + 1.195)$
6	$(s^2 + 0.856s + 0.263)(s^2 + 0.626s + 0.696)(s^2 + 0.229s + 1.129)$
7	$(s + 0.337)(s^2 + 0.679s + 0.330)(s^2 + 0.470s + 0.753)(s^2 + 0.168s + 1.069)$
8	$(s^2 + 0.643s + 0.146)(s^2 + 0.545s + 0.416)(s^2 + 0.364s + 0.779)(s^2 + 0.128s + 1.069)$

Chebyshev 0.5 dB

n

1	$(s + 2.863)$
2	$(s^2 + 1.426s + 1.516)$
3	$(s + 0.626)(s^2 + 0.626s + 1.142)$
4	$(s^2 + 0.847s + 0.356)(s^2 + 0.351s + 1.064)$
5	$(s + 0.362)(s^2 + 0.586s + 0.477)(s^2 + 0.224s + 1.036)$
6	$(s^2 + 0.580s + 0.157)(s^2 + 0.424s + 0.590)(s^2 + 0.155s + 1.023)$
7	$(s + 0.256)(s^2 + 0.462s + 0.254)(s^2 + 0.319s + 0.677)(s^2 + 0.114s + 1.016)$
8	$(s^2 + 0.439s + 0.088)(s^2 + 0.372s + 0.359)(s^2 + 0.248s + 0.741)(s^2 + 0.087s + 1.012)$

Chebyshev 1 dB

n

1	$(s + 1.965)$
2	$(s^2 + 1.098s + 1.103)$
3	$(s + 0.494)(s^2 + 0.494s + 0.994)$
4	$(s^2 + 0.674s + 0.279)(s^2 + 0.279s + 0.987)$
5	$(s + 0.289)(s^2 + 0.468s + 0.429)(s^2 + 0.179s + 0.988)$
6	$(s^2 + 0.464s + 0.125)(s^2 + 0.340s + 0.558)(s^2 + 0.124s + 0.991)$
7	$(s + 0.205)(s^2 + 0.370s + 0.230)(s^2 + 0.256s + 0.653)(s^2 + 0.091s + 0.993)$
8	$(s^2 + 0.352s + 0.070)(s^2 + 0.298s + 0.341)(s^2 + 0.199s + 0.724)(s^2 + 0.070s + 0.994)$

Chebyshev 2 dB

n

1	$(s + 1.308)$
2	$(s^2 + 0.804s + 0.823)$
3	$(s + 0.369)(s^2 + 0.369s + 0.886)$
4	$(s^2 + 0.506s + 0.222)(s^2 + 0.210s + 0.929)$
5	$(s + 0.218)(s^2 + 0.353s + 0.393)(s^2 + 0.135s + 0.952)$
6	$(s^2 + 0.351s + 0.100)(s^2 + 0.257s + 0.533)(s^2 + 0.094s + 0.966)$
7	$(s + 0.155)(s^2 + 0.280s + 0.212)(s^2 + 0.194s + 0.635)(s^2 + 0.069s + 0.975)$
8	$(s^2 + 0.266s + 0.057)(s^2 + 0.226s + 0.327)(s^2 + 0.151s + 0.710)(s^2 + 0.053s + 0.980)$

Chebyshev 3 dB

n

1     $(s+1.002)$
2     $(s^2+0.645s+0.708)$
3     $(s+0.299)(s^2+0.299s+0.839)$
4     $(s^2+0.411s+0.196)(s^2+0.170s+0.903)$
5     $(s+0.178)(s^2+0.287s+0.377)(s^2+0.110s+0.936)$
6     $(s^2+0.285s+0.089)(s^2+0.209s+0.522)(s^2+0.07s+0.955)$
7     $(s+0.126)(s^2+0.228s+0.204)(s^2+0.158s+0.627)(s^2+0.056s+0.966)$
8     $(s^2+0.217s+0.050)(s^2+0.184s+0.321)(s^2+0.123s+0.704)(s^2+0.043s+0.974)$

Bessel

n

1     $(s+1.000)$
2     $(s^2+3.000s+3.000)$
3     $(s+2.322)(s^2+3.678s+6.459)$
4     $(s^2+4.208s+11.488)(s^2+5.792s+9.140)$
5     $(s+3.647)(s^2+4.649s+18.156)(s^2+6.704s+14.272)$
6     $(s^2+5.032s+26.514)(s^2+7.471s+20.853)(s^2+8.497s+18.801)$
7     $(s+4.972)(s^2+5.371s+36.597)(s^2+8.140s+28.937)(s^2+9.517s+25.666)$
8     $(s^2+5.678s+48.432)(s^2+8.737s+38.569)(s^2+10.410s+33.935)(s^2+11.176s+31.977)$

Appendix D: Formulas of Normalized Filters

$$\omega_c = \omega_1 = \omega_2 = \omega_0 = 1 \quad \text{rad/s}$$

D.1 SALLEN–KEY FILTERS

(a) First-order $(K = 1)$

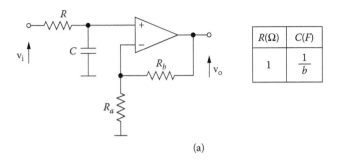

$R(\Omega)$	$C(F)$
1	$\dfrac{1}{b}$

(a)

(b) Second-order

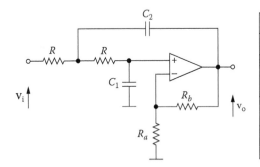

$R = 1\,\Omega,$	$R_a = 1\,\Omega,\quad R_b = K - 1\,\Omega$	
K	$C_1(f)$	$C_2(f)$
> 1	$\dfrac{a + \sqrt{a^2 + 8b\,(K - 1)}}{4b}$	$\dfrac{1}{bC_1}$
1	$\dfrac{a}{2b}$	$\dfrac{2}{a}$
$3 - \dfrac{a}{\sqrt{b}}$	$\dfrac{1}{\sqrt{b}}$	$\dfrac{1}{\sqrt{b}}$

(b)

D.2 MULTIFEEDBACK FILTERS

(a) Low-pass

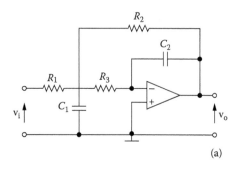

$R_1 = R_3 = R = 1\ \Omega$		
$R_2\ (\Omega)$	$C_1(F)$	$C_2\ (F)$
1	$\dfrac{2K+1}{aK}$	$\dfrac{a}{(2K+1)b}$

(a)

(b)   Band-pass

(i) BPF with one op-amp

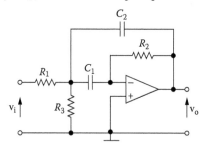

	$C = 1\ F$		$Q \le 10$
K	$R_1(\Omega)$	$R_2(\Omega)$	$R_3(\Omega)$
$\dfrac{R_2}{2R_1}$	$\dfrac{Q}{K}$	$2Q$	$\dfrac{Q}{2Q^2 - K}$
$2Q^2$	$\dfrac{Q}{K}$	$2Q$	∞

(ii) Deliyannis BPF

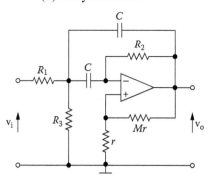

$C = 1\ F,$		$R_a = 1\ F$		$Q \le 30$
$R(\Omega)$	$R_1(\Omega)$	$R_2(\Omega)$	$R_3(\Omega)$	R_b
$\dfrac{1}{\sqrt{k}}$	KR	kR	$\dfrac{KR_n}{K-1}$	M
K: Free Parameter				
$\dfrac{1}{\sqrt{k}}$	R	kR	∞	M
$k = 2^2, 3^2, \dots\ M = \dfrac{kQ}{2Q - \sqrt{k}}$			$K = \dfrac{Q(1+M)\sqrt{k}}{M}$	

(iii) BPF with two op-amps

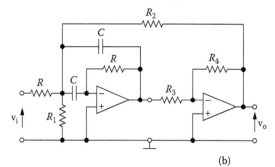

$C = 1\ F$	$R_3 = R_4 = 1\ \Omega$ $Q \le 50$
$R_1(\Omega)$	$R_2(\Omega)$
Q	$\dfrac{Q}{Q^2 - (Q+1)}$

(b)

(c)   Notch

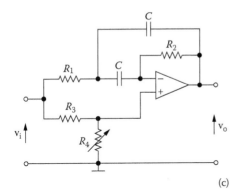

$C = 1\,F$	$R_3 = 1\,\Omega$	$Q \le 10$
$R_1(\Omega)$	$R_2(\Omega)$	$R_4(\Omega)$
$\dfrac{1}{2Q}$	$2Q$	$2Q^2$

(c)

D.3 FILTERS WITH 3 OP-AMPS

(a)   State-variable

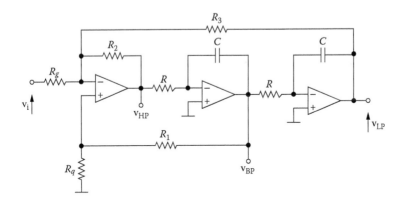

$C = 1\,F$	$R_g = R_q = R = 1\,\Omega$		$Q \le 100$		
	$R_1(\Omega)$	$R_2(\Omega)$	$R_3(\Omega)$		
LPF	$\dfrac{1 + b + bk}{a} - 1$	bK	K		
HPF	$\dfrac{1 + b + bk}{a} - 1$	K	$\dfrac{K}{b}$		
BPF	$3Q - 1$	1	1	$K = Q$ Free Parameter	

(a)

(b) Biquad filter

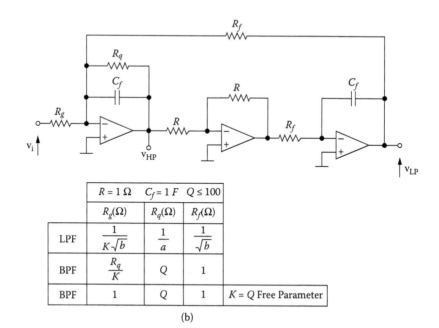

	$R = 1\,\Omega$	$C_f = 1\,F$	$Q \le 100$	
	$R_g(\Omega)$	$R_q(\Omega)$	$R_f(\Omega)$	
LPF	$\dfrac{1}{K\sqrt{b}}$	$\dfrac{1}{a}$	$\dfrac{1}{\sqrt{b}}$	
BPF	$\dfrac{R_q}{K}$	Q	1	
BPF	1	Q	1	$K = Q$ Free Parameter

(b)

Appendix E

TABLE E1
Element Values for Low-Pass LC Butterworth Filters

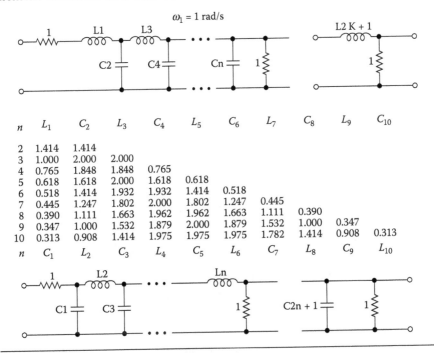

$\omega_1 = 1$ rad/s

n	L_1	C_2	L_3	C_4	L_5	C_6	L_7	C_8	L_9	C_{10}
2	1.414	1.414								
3	1.000	2.000	2.000							
4	0.765	1.848	1.848	0.765						
5	0.618	1.618	2.000	1.618	0.618					
6	0.518	1.414	1.932	1.932	1.414	0.518				
7	0.445	1.247	1.802	2.000	1.802	1.247	0.445			
8	0.390	1.111	1.663	1.962	1.962	1.663	1.111	0.390		
9	0.347	1.000	1.532	1.879	2.000	1.879	1.532	1.000	0.347	
10	0.313	0.908	1.414	1.975	1.975	1.975	1.782	1.414	0.908	0.313
n	C_1	L_2	C_3	L_4	C_5	L_6	C_7	L_8	C_9	L_{10}

TABLE E2
Element Values for Low-Pass Chebyshev Filters

Chebyshev 0.01 dB

n	L_1	C_2	L_3	C_4	L_5	C_6	L_7	C_8	L_9
3	1.181	1.821	1.181						
5	0.977	1.685	2.037	1.685	0.977				
7	0.913	1.595	2.002	1.870	2.002	1.595	0.913		
9	0.885	1.551	1.961	1.862	2.072	1.862	1.961	1.551	0.885
n	C_1	L_2	C_3	L_4	C_5	L_6	C_7	L_8	C_9

Chebyshev 0.1 dB

n	L_1	C_2	L_3	C_4	L_5	C_6	L_7	C_8	L_9
3	1.433	1.594	1.433						
5	1.301	1.556	2.241	1.556	1.301				
7	1.262	1.520	2.239	1.680	2.239	1.520	1.262		
9	1.245	1.502	2.222	1.683	2.296	1.683	2.222	1.502	1.245
n	C_1	L_2	C_3	L_4	C_5	L_6	C_7	L_8	C_9

Chebyshev 0.25 dB

n	L_1	C_2	L_3	C_4	L_5	C_6	L_7	C_8	L_9
3	1.633	1.436	1.633						
5	1.540	1.435	2.440	1.435	1.540				
7	1.512	1.417	2.453	1.535	2.453	1.417	1.512		
9	1.500	1.408	2.445	1.541	2.508	1.541	2.445	1.408	1.500
n	C_1	L_2	C_3	L_4	C_5	L_6	C_7	L_8	C_9

Chebyshev 0.5 dB

n	L_1	C_2	L_3	C_4	L_5	C_6	L_7	C_8	L_9
3	1.864	1.280	1.864						
5	1.807	1.302	2.691	1.302	1.807				
7	1.790	1.296	2.718	1.385	2.718	1.296	1.790		
9	1.782	1.292	2.716	1.392	2.773	1.392	2.716	1.292	1.782
n	C_1	L_2	C_3	L_4	C_5	L_6	C_7	L_8	C_9

Chebyshev 1 dB

n	L_1	C_2	L_3	C_4	L_5	C_6	L_7	C_8	L_9
3	2.216	1.088	2.216						
5	2.207	1.128	3.102	1.128	2.207				
7	2.204	1.131	3.147	1.194	3.147	1.131	2.204		
9	2.202	1.131	3.154	1.202	3.208	1.202	3.154	1.131	2.202
n	C_1	L_2	C_3	L_4	C_5	L_6	C_7	L_8	C_9

Chebyshev 2 dB

n	L_1	C_2	L_3	C_4	L_5	C_6	L_7	C_8	L_9
3	2.800	0.860	2.800						
5	2.864	0.909	3.827	0.909	2.864				
7	2.882	0.917	3.901	0.959	3.901	0.917	2.882		
9	2.890	0.920	3.920	0.968	3.974	0.968	3.920	0.920	2.890
n	C_1	L_2	C_3	L_4	C_5	L_6	C_7	L_8	C_9

Chebyshev 3 dB

n	L_1	C_2	L_3	C_4	L_5	C_6	L_7	C_8	L_9
3	3.350	0.712	3.350						
5	3.482	0.762	4.538	0.762	3.482				
7	3.519	0.772	4.639	0.804	4.639	0.772	3.519		
9	3.534	0.776	4.669	0.812	4.727	0.812	4.669	0.776	3.534
n	C_1	L_2	C_3	L_4	C_5	L_6	C_7	L_8	C_9

TABLE E3
Element Values for Elliptic Low-Pass Filters

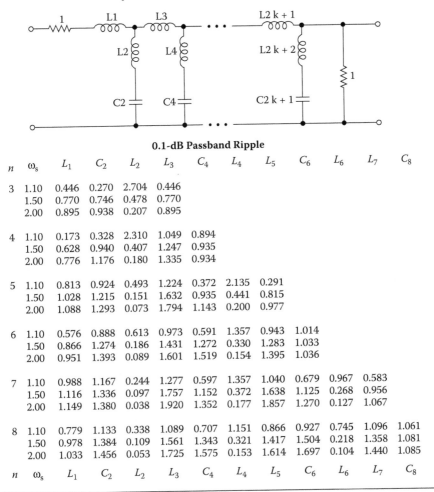

0.1-dB Passband Ripple

n	ω_s	L_1	C_2	L_2	L_3	C_4	L_4	L_5	C_6	L_6	L_7	C_8
3	1.10	0.446	0.270	2.704	0.446							
	1.50	0.770	0.746	0.478	0.770							
	2.00	0.895	0.938	0.207	0.895							
4	1.10	0.173	0.328	2.310	1.049	0.894						
	1.50	0.628	0.940	0.407	1.247	0.935						
	2.00	0.776	1.176	0.180	1.335	0.934						
5	1.10	0.813	0.924	0.493	1.224	0.372	2.135	0.291				
	1.50	1.028	1.215	0.151	1.632	0.935	0.441	0.815				
	2.00	1.088	1.293	0.073	1.794	1.143	0.200	0.977				
6	1.10	0.576	0.888	0.613	0.973	0.591	1.357	0.943	1.014			
	1.50	0.866	1.274	0.186	1.431	1.272	0.330	1.283	1.033			
	2.00	0.951	1.393	0.089	1.601	1.519	0.154	1.395	1.036			
7	1.10	0.988	1.167	0.244	1.277	0.597	1.357	1.040	0.679	0.967	0.583	
	1.50	1.116	1.336	0.097	1.757	1.152	0.372	1.638	1.125	0.268	0.956	
	2.00	1.149	1.380	0.038	1.920	1.352	0.177	1.857	1.270	0.127	1.067	
8	1.10	0.779	1.133	0.338	1.089	0.707	1.151	0.866	0.927	0.745	1.096	1.061
	1.50	0.978	1.384	0.109	1.561	1.343	0.321	1.417	1.504	0.218	1.358	1.081
	2.00	1.033	1.456	0.053	1.725	1.575	0.153	1.614	1.697	0.104	1.440	1.085
n	ω_s	L_1	C_2	L_2	L_3	C_4	L_4	L_5	C_6	L_6	L_7	C_8

1-dB Passband Ripple

n	w_s	L_1	C_2	L_2	L_3	C_4	L_4	L_5	C_6	L_6	L_7	C_8
3	1.10	1.225	0.375	1.948	1.225							
	1.50	1.692	0.733	0.486	1.692							
	2.00	1.852	0.859	0.226	1.852							
4	1.10	0.809	0.540	1.400	1.181	1.450						
	1.50	1.257	1.114	0.344	1.390	1.532						
	2.00	1.407	1.324	0.160	0.468	1.551						
5	1.10	1.697	0.775	0.588	1.799	0.399	1.989	1.121				
	1.50	1.977	0.977	0.188	2.492	0.794	0.520	1.719				
	2.00	2.056	1.034	0.092	2.736	0.936	0.249	1.919				
6	1.10	1.221	0.942	0.577	1.109	0.757	1.058	1.017	1.647			
	1.50	1.554	1.259	0.188	1.625	1.466	0.287	1.270	1.725			
	2.00	1.657	1.354	0.092	1.809	1.724	0.136	1.357	1.744			
7	1.10	1.910	0.927	0.307	1.936	0.480	1.688	1.553	0.593	1.107	1.420	
	1.50	2.079	1.048	0.100	2.614	0.874	0.490	2.440	0.905	0.333	1.877	
	2.00	2.124	1.080	0.049	2.844	1.016	0.235	2.753	1.006	0.160	2.019	
8	1.10	1.466	1.101	0.348	1.288	0.759	1.072	0.900	1.113	0.620	1.078	1.760
	1.50	1.719	1.283	0.118	1.837	1.369	0.315	1.438	1.772	0.185	1.261	1.830
	2.00	1.791	1.334	0.058	2.028	1.589	0.152	1.626	1.996	0.089	1.321	1.848
n	w_s	L_1	C_2	L_2	L_3	C_4	L_4	L_5	C_6	L_6	L_7	C_8

Appendix F: Coefficients of Denominator Polynomial

$$s^n + b_{n-1}s^{n-1} + \cdots + b_2s^2 + b_1s + b_o$$

TABLE F1
Butterworth Filters

n	b_0	b_1	b_2	b_3	b_4	b_5	b_6	b_7
1	1.000							
2	1.000	1.414						
3	1.000	2.000	2.000					
4	1.000	2.613	3.414	2.613				
5	1.000	3.236	5.236	5.236	3.236			
6	1.000	3.864	7.464	9.141	7.464	3.864		
7	1.000	4.494	10.098	14.592	14.592	10.098	4.494	
8	1.000	5.126	13.137	21.846	25.688	21.846	13.137	5.126

TABLE F2
0.1-dB Chebyshev Filter

n	b_0	b_1	b_2	b_3	b_4	b_5	b_6	b_7
1	6.552							
2	3.313	2.372						
3	1.638	2.630	1.939					
4	0.829	2.026	2.627	1.804				
5	0.410	1.436	2.397	2.771	1.744			
6	0.207	0.902	2.048	2.779	2.966	1.712		
7	0.102	0.562	1.483	2.705	3.169	3.184	1.693	
8	0.052	0.326	1.067	2.159	3.419	3.565	3.413	1.681

TABLE F3
0.5-dB Chebyshev Filter

n	b_0	b_1	b_2	b_3	b_4	b_5	b_6	b_7
1	2.863							
2	1.516	1.426						
3	0.716	1.535	1.253					

4	0.379	1.025	1.717	1.197				
5	0.179	0.753	1.310	1.937	1.172			
6	0.095	0.432	1.172	1.590	2.172	1.159		
7	0.045	0.282	0.756	1.648	1.869	2.413	1.151	
8	0.024	0.153	0.574	1.149	2.184	2.149	2.657	1.146

TABLE F4
1-dB Chebyshev Filter

n	b_0	b_1	b_2	b_3	b_4	b_5	b_6	b_7
1	1.965							
2	1.103	1.098						
3	0.491	1.238	0.988					
4	0.276	0.743	1.454	0.953				
5	0.123	0.581	0.974	1.689	0.937			
6	0.069	0.307	0.939	1.202	1.931	0.928		
7	0.031	0.214	0.549	1.358	1.429	2.176	0.923	
8	0.017	0.107	0.448	0.847	1.837	1.655	2.423	0.920

TABLE F5
2-dB Chebyshev Filter

n	b_0	b_1	b_2	b_3	b_4	b_5	b_6	b_7
1	1.308							
2	0.823	0.804						
3	0.327	1.022	0.738					
4	0.206	0.517	1.256	0.716				
5	0.082	0.459	0.693	1.450	0.706			
6	0.051	0.210	0.771	0.867	1.746	0.701		
7	0.020	0.166	0.383	1.144	1.039	1.994	0.698	
8	0.013	0.070	0.360	0.598	1.580	1.212	2.242	0.696

TABLE F6
3-dB Chebyshev Filter

n	b_0	b_1	b_2	b_3	b_4	b_5	b_6	b_7
1	1.002							
2	0.708	0.645						
3	0.251	0.928	0.597					
4	0.177	0.405	1.169	0.582				
5	0.063	0.408	0.549	1.415	0.574			
6	0.044	0.163	0.699	0.691	1.663	0.571		
7	0.016	0.146	0.300	1.052	0.831	1.912	0.568	
8	0.011	0.056	0.321	0.472	1.467	0.972	2.161	0.567

TABLE F7
Bessel Filter

n	b_0	b_1	b_2	b_3	b_4	b_5	b_6
1	1						
2	3	3					
3	15	15	6				
4	105	105	45	10			
5	945	945	420	105	15		
6	10395	10395	4725	1260	210	21	
7	135135	135135	62370	17325	3150	378	28

Bibliography

1. Acar, C., Anday, F., and Kutman, H., On the realization of OTA-C filters, *Int. J. Circuit Theory Applications,* 21(4), 331–341, 1993.
2. Allen, P. E. and Sanchez-Sinencio, E., *Switched Capacitor Circuits*, Van Nostrand Reinhold, New York, 1984.
3. Berlin, H. M., *Design of Active Filters*, H. W. Sams, New York, 1981.
4. Bruton, L. T., *RC-Active Circuits*, Prentice-Hall, Englewood Cliffs, New Jersey, 1981.
5. Budak, A., *Passive and Active Network Analysis and Synthesis*, Hougston, Mifflin, 1974.
6. Chen, C., *Active Filter Design*, Hayden, 1982.
7. Christian, E., *LC Filters*, John Wiley and Sons, New York, 1983.
8. Clayton, G. B., *Linear Applications Handbook*, TAB Books, 1975.
9. Daniels, R. W., *Approximation Method of Electronic Filter Design*, McGraw-Hill, New York, 1974.
10. Davis, TH. W. and Palmer, R. W., *Computer-Aided Analysis of Electrical Networks*, C. E. Merrill Publ. Co., Columbus, OH, 1973, chap. 7.
11. Daryanani, G., *Principle of Active Network Synthesis and Design*, John Wiley & Sons, 1976.
12. Deliyannis, T., High-Q Factor Circuit with Reduced Sensitivity, *Electronics Letters*, Vol. 4, No. 26, 1968.
13. Deliyannis, T., Sun, Y., and Fidler, J. K., *Continuous-Time Active Filter Design*, CRC Press, Boca Raton, FL, 1999.
14. DePian, L., *Linear Active Network Theory*, Prentice-Hall, Englewood Cliffs, NJ, 1962.
15. Floyd, T., *Electronic Devices: Conventional-Flow Version*, 4th ed., Prentice-Hall, Englewood Cliffs, NJ, 1995.
16. Geiger, R. L. and Sanchez-Sinencio, E., Active Filters Design Using Operational Transconductance Ampifiers: A Tutorial, *IEEE, Circuit and Devices*, Vol. 1, pp. 20–32, 1985.
17. Geffe, P. R., *Simplified Modern Filter Design*, J. Rider, 1963.
18. Ghausi, M. S. and Laker, K. R., *Modern Filter Design*, Prentice Hall, 1981.
19. Graeme, J. D., Tobey, G. E., and Huelsman, L. P. (Eds.), *Operational Amplifiers: Design and Applications*, McGraw-Hill, New York, 1971 (Burr-Brown).
20. Grebene, A. B., *Bipolar and MOS Analog Integrated Circuit Design*, John Wiley & Sons, New York, 1984.
21. Gregorian, R. and Tems, G. C., *Analog MOS Integrated Signal Processing*, John Wiley & Sons, New York, 1986.
22. Guillemin, E. A., *Introductory Circuit Theory*, John Wiley & Sons, New York, 1953.
23. Hilburn, J. L. and Johnson, D. E., *Rapid Practical Design of Active Filters*, John Wiley & Sons, New York, 1975.
24. Huelsman, L. P., *Theory and Design of RC Active Circuits*, TMH edition, McGraw-Hill, New York, 1968.
25. Huelsman, L. P., *Active Filters: Lumped, Distributed, Integrated, Digital & Parametric*, McGraw-Hill, New York, 1970.
26. Huelsman, L. P. and Allen, P. E., *Introduction to the Theory and Design of Active Filters*, McGraw-Hill, New York, 1980.

27. Kincaid, R. and Shirley, F., Active Bandpass Filter Design Is Made Easy with Computer Program, *Electronics*, May 16, 1974.
28. Kuo, F. F., *Network Analysis and Synthesis*, John Wiley & Sons, New York, 1966.
29. Lacanette, K. (Ed.), *Switched-Capacitor Filter Handbook*, Natonal Semiconductor Corp., Santa Clara, 1985.
30. Lancaster, D., *Active-Filter Cookbook*, Sams, 1975.
31. Mitra, S. K., *Analysis and Synthesis of Linear Active Network*, John Wiley & Sons, New York, 1969.
32. Moberg, G. O., Multiple-Feedback Filter Has Low Q and High Gain, *Electronics*, December 9, 1976, pp. 97–99.
33. Moshytz, G. S. and Horn, P., *Active Filter Design Handbook*, John Wiley & Sons, New York, 1981.
34. Moshytz, G. S., Inductorless Filters: A Survey: Part III: Linear, Active & Digital Filters, *IEEE Spectrum*, September 1970, pp. 63–76.
35. Pactitis, S. A. and Kossidas, A. T., Design of Sallen-Key Active Filters, IASTED, Intern. Symp., August 1983.
36. Pactitis, S. A. and Kossidas, A. T., Computer Implementation of VCVS Active Filters, *Advance in System Modeling and Simulation*, North Holland, Amsterdam, 1989.
37. Peyton, A. J. and Walsh, V., *Analog Electronics with Op Amps*, Cambridge University Press, New York, 1993.
38. Russel, H. T., Design Active Filters with Less Effort, *Electronic Design*, January 7, 1971, pp. 82–85.
39. Sallen, R. P. and Key, E. L., A practical Method of Design RC Active Filters, *Trans. IRE*, CT-2, pp. 74–85, 1955.
40. Schaumann, R. and Van Valkenburg, M. E., *Design of Analog Filters*, Oxford University Press, New York, 2000.
41. Schaumann, R., Continous-time integrated filters, in *The Circuit and Filters Handbook*, Chen, W. K., Ed., CRC Press, Boca Raton, FL, 1995.
42. Schaumann, R., Ghausi, M. S., and Laker, K. R., *Design of Analog Filters, Passive, Active RC and Switching Capacitors*, Prentice-Hall, Englewood Cliffs, NJ, 1990.
43. Sedra, A. S. and Brakett, P., *Filter Theory and Design: Active and Passive*, Matrix Publishers, London, 1978.
44. Sun, Y. and Fidler, J. K., Novel OTA-C realization of biquadratic transfer functions, *Int. J. Electronics,* 75, 333–348, 1993.
45. Temes, G. C. and LaPatra, S. K., *Circuit Synthesis and Design*, McGraw-Hill, New York, 1977.
46. Tobey, Graeme, and Huelsman, *Operational Amplifier Design and Applications*, McGraw-Hill, New York, 1971.
47. Tomlison, G. H., *Electrical Networks and Filters*, Prentice Hall International, London, 1991.
48. Toumazou, C., Lidgey, F. J., and Haigh, D. G. (Eds.), *Analogue IC Design: The Current Mode Approach*, Peter Peregrinous, 1990.
49. Van Valkenburg, M. E., *Analog Filter Design*, Holt, Rinehart & Winston, New York, 1982.
50. Van Valkenburg, M. E., *Modern Network Synthesis*, John Wiley & Sons, New York, 1962.
51. Volpe, G. T. and Premisler, L., Universal Building Blocks Simplify Active Filter Design, *EDN*, September 5, 1976, pp. 91–95.
52. Chen, W. -K., *Passive and Active Filters, Theory and Implementations*, John Wiley & Sons, New York, 1986.
53. Williams, A. B., *Electronic Filter Design Handbook*, McGraw-Hill, New York, 1981.

Index

Lightning Source UK Ltd.
Milton Keynes UK
UKHW022317070223
416660UK00019B/108